The Mathematics That

P wer

Our World

How Is It Made?

The Mathematics That Power Our World

How Is It Made?

Joseph Khoury
Gilles Lamothe
University of Ottawa, Canada

World Scientific

NEW JERSEY · LONDON · SINGAPORE · BEIJING · SHANGHAI · HONG KONG · TAIPEI · CHENNAI · TOKYO

Published by

World Scientific Publishing Co. Pte. Ltd.

5 Toh Tuck Link, Singapore 596224

USA office: 27 Warren Street, Suite 401-402, Hackensack, NJ 07601

UK office: 57 Shelton Street, Covent Garden, London WC2H 9HE

British Library Cataloguing-in-Publication Data
A catalogue record for this book is available from the British Library.

THE MATHEMATICS THAT POWER OUR WORLD
How Is It Made?

ISBN 978-981-4730-84-6
ISBN 978-981-3144-08-8 (pbk)

Printed in Singapore

"To the person who made me watch dozens of the *How it's made* episodes during his childhood years, who continues to inspire me everyday with his never-ending curiosity. To the better version of me, my son Michel."

Joseph Khoury

Preface

In the early 21st century, a TV show called *How it's made* premiered in North America and quickly grew in popularity among viewers of all ages. The purpose of the show was to look behind the scenes to explain in simple terms how common everyday items are actually made. Although many episodes of the show featured simple things like the jeans we wear, the bicycle we ride or even some of the processed food we eat, it was certainly an eye opener on the ingenuity and effort behind the simplest things we use on a daily basis.

Unlike other scientists, many mathematicians often work relentlessly on solving difficult theoretical problems without questioning much the practical implications of their discoveries on the day-to-day life. This can be explained from cultural and practical points of view. On a cultural level, mathematics were perceived throughout human history as a form of collaborative art that can mark a whole nation and contribute as a measure of achievement of that nation. On a practical level, who is to say that the seemingly abstract problems that we are solving today are not the fuel of the new technology in the future? In 1940, G.H. Hardy, a mathematician known for his achievements in number theory and mathematical analysis, wrote an essay called *A Mathematician's Apology*. He described the most beautiful mathematics as those without real-life applications. In fact, he described number theory as "useless", yet very elegant and beautiful. His motive was (in part at least) to promote the idea that mathematics should be pursued for its own beauty and not for the sake of its applications. Of course, it was hard in the early 40's of the last century to imagine that number theory would become an important tool in the field of modern cryptography. Public-key encryption (and thus internet banking) is one application, among many others, that would not be possible without the great accomplishment of number theorists such as Hardy. Regardless how we value the research in mathematics, the fact remains that a wide variety of phenomena around us are governed by mathematical models. Pushing for innovation and research in applied and pure mathematics is key to make further advancement in science, technology and even medicine. To say the least, mathematics are certainly a set of tools, among many others, that help make modern technology function well.

Why did we write the book?

In this technology-driven era, our modern lifestyle is certainly not short of devices that we use on a daily basis regardless of age, gender, language or socio-political backgrounds. Whether receiving or sending a text message, taking a digital photo, undergoing an MRI scan or following your GPS instructions blindly to get to your destination, you are benefiting from the collaborative effort of scientists and engineers. The rapid pace of advancements in technology makes it so hard on most of us to keep track of what is new, let alone to pause and learn about the science behind it.

While embracing the seemingly never ending advances in technology, you must have wondered at one point: How does the GPS know where you are? What magic makes the Google search engine classify and display the results of your search query? Even the simplest of all: How does your pocket calculator display, in a fraction of a second, the answer to a complicated arithmetic operation? To most people, mathematics do not come up as a possible answer to these questions, while in fact, they are the driving force behind all of them. A major difficulty encountered in high school and university mathematics courses for students registered in programs other than pure mathematics is the disconnect between what they learn in their mathematics courses and the relevance of this material with their chosen fields. This can seriously affect their motivation and their ultimate success. In fact, a question instructors of mathematics hear very often from their students is: *Why do we need to learn this*? Things can get even worst when it comes to research in mathematics. On various social encounters, mathematicians are often greeted with statements like *"Math was my worst subject"*, and *"Hasn't everything been discovered in mathematics?"* or *"What is the latest number that you discovered?"* Most people associate mathematicians with pure academia, but very few are actually aware that there are many mathematicians working in industry, national security and even the medical field. In fact, many of the best paying jobs for new university graduates require a strong mathematical background. For example, a software developer, an investment banker, an actuary, an engineer or a financial analyst.

With the hope of convincing students that there is a need to acquire mathematical skills, and to introduce the general public to the pivotal role played by mathematics in our lives, we started our endeavor of "looking under the hood" at the engine that makes most of the technology around us run smoothly. But there is another, stronger motivation for starting this book. After years of teaching various areas of undergraduate mathematics, we realized that the traditional place held by mathematics in education for many centuries is taking a step backward. In the name of reform and adapting to a fast changing world, the learning of mathematics has unfortunately degraded in many cases into an empty drill of memorization of miscellaneous techniques rather than a foundation of scientific reasoning critical in any aspect of knowledge. More and more, the mathematical community seems to be divided into two extremes. On one extreme, we have mathematicians who disassociate their teaching almost completely from any aspect of scientific thinking. They transfer their knowledge of the subject matter in the form of "recipes" for their students to follow almost blindly. On the other extreme, we have the group of mathematicians with an overemphasis on abstraction and almost a complete disconnection from real applications even in early service undergraduate courses in mathematics. Our hope is that this book will contribute as a middle ground between the two extremes.

Who is the intended audience for the book?

The topics for the five chapters of the book are carefully chosen to strike a delicate balance between relevant common applications and a reasonable dose of mathematics. In most chapters, the mathematical maturity needed is acquired after a year of studies at the university level in any branch of science or engineering. However, self-motivated advanced high school students with a strong desire to acquire more knowledge and a willingness to expand their horizons beyond the school curriculum can certainly benefit a great deal from researching and understanding the mathematics in the book. The topics discussed in the book are also great resources for high school teachers and university professors who can use the various applications to go hand-in-hand with the theory taught in class.

Organization of the book

Throughout the book, all efforts were made to keep the mathematical requirement to a minimum. For some advanced topics, like the theory of finite fields and the notion of primitive polynomials in Chapter 4, the mathematical requirement is made in an almost self-contained fashion. All the topics are presented with a fair amount of details. Chapters are independent to a large extent. The readers can choose a topic to read without acquiring full knowledge of previous chapters. At the beginning of each chapter, a small section entitled *Before you go further* is included to give the reader an idea about the level of mathematical knowledge required to fully understand the chapter. The organization of the chapters in the book are as follows.

Chapter 1 discusses the mathematics of an electronic calculator. It starts with a review of basic number systems and their properties. Signed numbers and their digital representations, in particular the one's and two's complement schemes, are explained. Logic gates, which are at the heart of any digital computing, are introduced and a link to Boolean Algebra is made. Binary adders and adder-subtractor circuits are studied from the point of view of Boolean Algebra rules. The famous seven-segment display is explained.

Chapter 2 discusses the well-known Huffman codes, an essential tool in many data compression techniques. The chapter also provides an introduction to data compression and its modes (lossy and lossless). Binary codes, binary trees, uniquely decodable and prefix-free codes are introduced as well as Kraft's inequality and its applications. A brief introduction to information theory via the notion of entropy is given. The chapter ends with the Huffman algorithm to solve the famous optimal prefix-free binary code problem. Detailed examples are given.

Chapter 3 describes the JPEG standard. We all have uploaded or downloaded a picture with ".jpeg" extension at some point. The aim of the chapter is to explain how, this popular compression technique, is capable of storing or viewing photos or texts on your machine (computer or digital camera) with a significantly reduced storage size without jeopardizing the quality of the original file. The main tool for this technique is a transformation called the Discrete Cosine Transform. Concepts from linear algebra like matrix manipulations, linear independence and orthogonal bases are used. Some knowledge of basic properties of trigonometric functions and complex numbers is also required.

Chapter 4 is devoted to the study of the GPS system both from the satellite and the receiver ends. Although the mathematics used by the receiver to locate positions on the surface of the planet is fairly simple, the nature of the signals emitted by the satellites and the way the receiver interprets and treats them require heavy mathematics. Mathematical preliminaries necessary to understand the signal structure include group theory, modular arithmetics, the finite field $\mathbb{F}_p$, polynomial rings over $\mathbb{F}_p$ and the notion of primitive polynomials. This chapter is certainly the richest in terms of mathematical knowledge. If you have a curious mind and enjoy new challenges, this is definitely a chapter for you.

Chapter 5 discusses the manipulation of digital images, in particular we show the reader how to produce an "average" face by combining the digital images of many faces. This procedure is one of the steps involved for face recognition. In the last twenty years, face recognition has become a popular area of research in computer vision. We discuss one of the methods that have been proposed for this application, which is known as principal components analysis. Face recognition is an essential tool in security and forensic investigation in our modern society. Concepts from linear algebra like the dot product and orthogonal projections are used. A bit of knowledge in basic statistical concepts is also useful for fully understanding this chapter.

A word of caution

While every attempt is made to make every chapter in the book as complete as possible, some technical details are omitted as we are not experts in the specific domain of application nor do we claim to be. Technicalities like the way a circuit is wired, the type of transistors needed for a particular design or the nature of the electrical pulse of a satellite signal are beyond the scope of this book. Interested readers are encouraged to look up these aspects in books written by experts in the domain of application.

Contents

Chapter 1

What makes a calculator calculate?

1.1 Introduction

Throughout history, many civilizations realized the need to invent "counting machines" to help with long and complex calculations. Some forms of calculators were known even before number systems were fully developed. Early calculators were mainly mechanical using parts like levers, gears, axles and rods. The development of new mathematical counting algorithms paved the way to new types of counting machines to appear in Europe in the 17th century. But it was not until the early 1960's that the real revolution in electronic calculators took place thanks to the invention of a device called the *transistor*.

With the recent advancements in technology, you can hardly avoid seeing an electronic calculator around you as there is one built in almost every device you own: your phone, computer, tablet or even in your hand watch. We trust them blindly in our everyday tasks without questioning the answers they display. But have you ever wondered, "How can your pocket calculator do complex mathematical operations with such a high precision in a blink of an eye?" If you do not know the answer to that, you are certainly not alone. Ask your friends, even your math and science instructors and you will be surprised how little is known about the basics of this electronic device. This chapter takes you on a journey to explore some of the logic that powers digital computing.

1.1.1   *A view from the inside*

So what is inside an electronic calculator that makes it do its magic? Look around your house, you most likely can find an old or a broken calculator. If you were to take it apart (if you have not done that already once in your life), you will be surprised how little you will find inside. The heart of a typical electronic calculator is a microchip called the *processor*. The rest consists mainly of plastic keys for your inputs placed on top of a rubber membrane covering a certain touch activated electronic circuit. Calculators are also equipped with a certain form of display screen and power source like a lithium battery. The main focus of this chapter is, however, not so much on the hardware of the calculator but rather on the mathematics behind its operation that make all the pieces come together and work as a unit. We will not touch on the practical aspects or types of electrical or electronic parts used like transistors and resistors. Interested readers can definitely learn more about that side of the calculator from any book on basic electronics.

1.1.2   *Before you go further*

This chapter is probably the lightest in the book in terms of mathematical requirement. Most of the mathematics needed is explained with a reasonable amount of details. However, some mathematical maturity, reasoning and basic manipulation skills are required.

1.2   Number systems

I still have a picture in my head of one of my school teachers writing on the board the following expansion of the decimal number 1234:

$$1234 = \left(1 \times 10^3\right) + \left(2 \times 10^2\right) + \left(3 \times 10^1\right) + \left(4 \times 10^0\right)$$

and then saying that the *multiplier* (or coefficient) of 10^0 (4 in this example) is called the units digit, that of 10^1 (3 in this example) is called the tens digit and the other two multipliers (2 and 1) are called the hundreds and the thousands digit respectively. One thing the teacher did not explain at that time is why we chose powers of 10 in the above expansion. With time, I came to realize that there is really nothing special about the number 10 aside from the fact that humans have 10 fingers and 10 toes and that we tend to use them as we count. At least, this is what many historians seem to agree on as the main reason for using 10 as a base for our number system. This number system, familiar to all of us, is known as the *decimal system*. In it, every number is written using the 10 digits 0 to 9.

Given an integer $b \geq 2$, we can talk about the *number system with base b* in a similar way to our decimal system. In such a system, every number can be written using the digits from the set $S = \{0, 1, 2, \ldots, b-1\}$. More precisely, if $N = N_{k-1}N_{k-2} \ldots N_1 N_0$ (where each $N_i \in S$) is a number in the system then N holds the following decimal value:

$$N_{k-1}b^{k-1} + N_{k-2}b^{k-2} + \cdots + N_1 b + N_0 b^0.$$

For the number $N = N_{k-1}N_{k-2} \ldots N_1 N_0$, the digits N_0 and N_{k-1} are called the least significant digit (LSD) and the most significant digit (MSD) respectively. If $c > b$, then a number N written in base b can clearly be interpreted as a number in base c as well. This confusion could be avoided by specifying the base as a subscript and write $(N)_b$. Hence, $(123)_4 = (1 \times 4^2) + (2 \times 4^1) + (3 \times 4^0) = 27$ and $(123)_6 = (1 \times 6^2) + (2 \times 6^1) + (3 \times 6^0) = 51$.

Two digital representations of numbers are of particular importance for us in this chapter, the *binary system* and the *Binary coded decimal representation* (BCD for short). The first one is the language of every modern digital device and the second is used because of its ability to easily decipher coded data to its original form. Both systems transform any decimal number into a string of 0's and 1's.

1.2.1   *Why 0's and 1's?*

It is safe to say that the most common thing known about computers among non experts is the fact that "they use 0's and 1's". Before we go on to study the two digital representations, maybe it is now the right place to quickly address the question, "What does it really mean that a computer uses strings of 0's and 1's?" A short (but incomplete) answer is the following. Electronic devices are built using a number of chips each containing a fairly large number of transistors. In the context of this chapter, you can think of a transistor as a switch that you turn on or off as you press various keys on your device, just like your room light switches. An electrical current passing through a transistor is indicated by 1 (high voltage) and no current

Table 1.1   Division by 2.

Division by 2	Quotient	Remainder
165	82	1
82	41	0
41	20	1
20	10	0
10	5	0
5	2	1
2	1	0
1	0	1

is indicated by 0 (low voltage). By turning on and off these switches, we actually have a way to communicate with computers and give them instructions. A basic example is the representations of numbers. As you will see in the next section, a computer using an 8-bit system interprets the integer 8 as being the string **00001000**. This means that when you press the key labeled 8 on the keypad of your device, you are actually sending an electrical signal to a certain chip in your computer instructing it to turn off the first four of its switches, turn on the fifth switch and again off the last three switches. The processor of your device will interpret this series of switches as the number 8 and gets ready to operate on it based on your next action.

1.2.2   *The binary system*

This is the number system with base 2. In other words, every number in this system is built out of two digits 0 and 1 that we call *bits*. A number in this system is called a *binary number* and has the form $N = N_{k-1}N_{k-2}\ldots N_1 N_0$ where each N_i is either 0 or 1. Because of the vital role the binary system plays in electronics, it is important to develop the ability of switching back and forth between decimal and binary systems with ease. Two main algorithms to achieve that task are the *successive division by 2* and the *sum of weights*. As the name suggests, the idea of the first algorithm is to successfully divide a given decimal number n by 2 and record the successive remainders (0 or 1) until we hit a quotient of zero. The list of remainders read from bottom to top would be the binary representation of the decimal number. As a consequence, the last remainder would be the most significant bit. For example, to convert the decimal number 165 to binary we record the successive remainders upon its division by 2 as shown in the Table 1.1.

Reading the column of remainders from bottom to top gives the following binary representation of 165: **10100101**.

The second algorithm works well for relatively small decimal numbers. We start by displaying the first 15 powers of 2:

$$2^0 = 1, \quad 2^1 = 2, \quad 2^2 = 4, \quad 2^3 = 8, \quad 2^4 = 16$$
$$2^5 = 32, \quad 2^6 = 64, \quad 2^7 = 128, \quad 2^8 = 256, \quad 2^9 = 512$$
$$2^{10} = 1024, \, 2^{11} = 2048, \, 2^{12} = 4096, \, 2^{13} = 8192, \, 2^{14} = 16348.$$

Given a decimal number n, we look for the largest integer r_1 such that $2^{r_1} \leq n$ and we let $n_1 = n - 2^{r_1}$. Again, look for the largest integer r_2 such that $2^{r_2} \leq n_1$ and let $n_2 = n_1 - 2^{r_2}$. Repeat this process for all successive values n_i's until you hit a certain k with $n_k = 0$. Starting with the largest power of 2 appearing in the above process, we record 1 as a multiplier of 2^j if 2^j is used and 0 if 2^j is not used in the process. If 2^l is the largest power of 2 appearing in the above process, then this method gives a binary representation with $l + 1$ bits. Let us look at an example by revisiting the decimal number 165 treated in the first method. The largest power of 2 less than or equal to 165 is 7 since $2^7 = 128$. Subtracting 128 from 165 gives 37. The largest power of 2 not exceeding 37 is 5 since $2^5 = 32$. Now $37 - 32 = 5$ and the largest power of 2 less than or equal to 5 is 2 since $2^2 = 4$. Finally $5 - 4 = 1$ and $2^0 - 1 = 0$. The powers of 2 appearing in the above process are 2^7, 2^5, 2^2 and 2^0. Therefore

$$165 = 1 \times 2^7 + 0 \times 2^6 + 1 \times 2^5 + 0 \times 2^4 + 0 \times 2^3 + 1 \times 2^2 + 0 \times 2^1 + 1 \times 2^0$$

and the decimal representation of 165 is the following 8-bit number: 10100101. Note that we have 8 bits in the binary representation of 165 since the largest power of 2 used is 2^7.

Converting an n-bit binary number to its decimal form is straightforward. Just use expansion of the binary number in terms of powers of 2. For instance, the decimal value of the binary number 1110001101 is

$$1 \times 2^9 + 1 \times 2^8 + 1 \times 2^7 + 0 \times 2^6 + 0 \times 2^5 + 0 \times 2^4 + 1 \times 2^3 + 1 \times 2^2 + 0 \times 2^1 + 1 \times 2^0 = 909.$$

Remark 1.1. An important thing to keep in mind is that computers work with fixed size storage. If a computer uses n-bit storage size, then each number is stored using n bits. For example, in a 4-bit machine, the integer 5 is stored as 00101. The same integer is stored as 00000101 in an 8-bit machine. Modern computers use 32-bit or 64-bit storage size.

Now for the first result of the chapter.

Proposition 1.1. In an n-bit machine, the range of (non-negative) decimal number that can be represented is $[0, 2^n - 1]$.

Proof.   Note that there is a total of 2^n different n-bit binary numbers since each bit can take only two possible values 0 and 1. The binary number $00\cdots00$ has a decimal value of zero and the largest positive number that can be represented in n-bit binary is $11\cdots11$ (n ones) which has a decimal value of $2^{n-1}+2^{n-2}+\cdots+2^1+2^0$. Note that this last sum can be rearranged as $1 + 2 + \cdots + 2^{n-1}$ which is a *finite geometric sum* containing n terms with 1 as the first term and the ratio of two consecutive terms is 2. The value of this sum is known to be $\frac{2^n-1}{2-1} = 2^n - 1$. $\square$

1.2.3   *Binary Coded Decimal representation (BCD)*

This is one of the earliest digital representation systems of decimal numbers. The BCD system has almost disappeared in modern computer designs but it is still in use in devices like your pocket calculator. In this system, digits in the decimal system (0 to 9) are represented by their 4-bit binary representations. The following table gives the BCD codes for the digits 0 through 9.

Decimal	BCD
0	0000
1	0001
2	0010
3	0011
4	0100
5	0101
6	0110
7	0111
8	1000
9	1001

Concatenation is then used to represent any decimal number in BCD. For example, the decimal number 427 is represented by 010000100111 in BCD:

$$\underbrace{0100}_{4}\,\underbrace{0010}_{2}\,\underbrace{0111}_{7}.$$

Note that with 4 bits, one can form $2^4 = 16$ different binary codes which means that 6 codes are not used in the BCD system. The binary codes 1010 (number 10 in decimal) through 1111 (number 15 in decimal) are considered as invalid codes and cannot be used in a digital design operating on BCD system.

There is more than one form of a BCD representation in the literature. The form presented above is called the Natural Binary Coded Decimal and it is the most straightforward one. The easy conversion between decimal and BCD is the main virtue of this representation. As we will see later, this ease of conversion comes in handy in displaying decimal numbers in digital devices. Main drawbacks of BCD representation are its inefficiency in terms of data usage and the complexity of circuit implementation.

1.2.4 *Signed versus unsigned binary numbers*

When talking about *integers* in the decimal system, the term includes both non-negative and negative integers. In Section 1.2.2 on the binary system, we have actually defined the binary form of non-negative integers only, called *unsigned binary integers*. In practice, it is crucial that a distinction between positive and negative numbers is made in any number system and a machine that cannot deal with negative numbers is practically useless. For a "pencil and paper" arithmetics, this distinction is simply made by the "+" sign for positive numbers and "−" sign for negative ones. In a calculator however, every piece of data is represented in a binary form because of hardware limitations. It is therefore of great importance to understand the techniques used to represent signed numbers in binary form. Three such techniques are presented in what follows.

1.2.4.1 *Sign-magnitude format*

An intuitive way to represent *signed* integers in a digital format is to dedicate the leftmost digit as a "sign digit". This is called the *sign-magnitude format*. In this format, a leftmost bit of 0 represents a positive number and a leftmost bit of 1 represents a negative number. The remaining bits of the binary number represent its magnitude (or absolute value). For example, to represent -93 in a sign-magnitude format in a 8-bit machine, we start by representing the absolute value $|-93| = 93$ in binary form with seven bits: 1011101. We add 1 as a leftmost bit to indicate that the number is indeed negative and we get 11011101.

Proposition 1.1 above shows that in an n-bit machine, the decimal range of unsigned binary numbers is from 0 to $2^n - 1$. In the sign-magnitude format, one bit (leftmost) is used as a sign which means that in this format the decimal range is $\left[-2^{n-1} - 1, \ 2^{n-1} - 1\right]$. The bad news is that zero has two possible representations: $100 \cdots 00$ (which represents $+0$) and $000 \cdots 00$ (which represents -0).

1.2.4.2 *The one's complement representation*

For a binary number N, the one's complement of N is simply the binary number obtained by converting each 0 to 1 and each 1 to 0 in N. For example, the one's complement of the binary number 00101101 is 11010010. In the one's complement format, a non-negative decimal has the same representation as in the sign-magnitude format (or just the n-binary format), but the representation of a negative decimal is different. For a negative decimal k, the one's complement representation is obtained by writing the one's complement of the binary form of $|k|$. For example, to get the one's complement representation of -23 in a 8-bit machine, we start by writing the 8-bit binary of 23: $(23)_{10} = (00010110)_2$ and then we flip the digits: 11101001. So, $(-23)_{10} = (11101001)_{\text{one's complement}}$.

As in the sign-magnitude format, 0 has two differences representations in this format, namely $00 \cdots 00$ and $11 \cdots 11$.

1.2.4.3 *The two's complement representation*

With the limitations of the sign-magnitude and the one's complement formats, modern computers are programmed to use different schemes to represent signed integers. One scheme that proved to be very friendly in terms of hardware and circuit design inside the machine is the *two's complement representation*. To explain the main idea behind this scheme, let us consider the following scenario. Imagine you want to watch a movie on your electronic device with a basic four-digit counter (in seconds) that starts to run from the reading 0000 as soon as you press the play button.

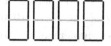

Assuming the movie running time is long enough (and you do not stop it), the counter will eventually read 9999. One second after that, it goes back to 0000. Now, imagine at this moment you hit the stop button and then rewind the movie for 5 seconds. The counter will probably read 9995.

Clearly, the reading '9995' in this case does not mean that the movie has been playing for 9995 seconds. To avoid the ambiguity on what 9995 means on the counter, one has to interpret the range 0 to 9999 a little differently. Note that $9995 + 5 = 10000 = 10^4$ and since the counter can only handle four digits, the leftmost bit (1 in this case) is dropped and we get 0000. This suggests that 9995 can be interpreted as -5 in this scenario since $9995 + 5$ results in 0000 displayed on the counter, exactly like $(-5) + 5 = 0$.

The above analogy with a digital counter was made to justify the following definition. Given an n-bit binary number N, the *two's complement* of N is defined to be the n-bit binary number $N_{2's}$ satisfying $N + N_{2's} = 1\underbrace{0\cdots0}_{n}$, with the "+" sign referring to binary addition that we will discuss in Section 1.3.1. Notice the analogy with the equation $9995 + 5 = 10000 = 10^4$ in the counter example above and the fact that the binary form of 2^n is precisely $10\cdots0$ (1 followed by n 0's). In other words, finding the two's complements of a binary number means finding the opposite (or negative) of this number.

The two's complement of a binary number N is obtained by finding first the one's complement of N and then adding 1 to the result. Practically, we have the following simple algorithm: read the digits of N from right to left outputting 0 as long as the digit is 0. Also output 1 for the first 1 you read. After that, invert the individual remaining bits of N, that is a 0 becomes a 1 and vice versa. For example, to find the two's complement of the binary number 10101000, start reading the digits from the right keeping the first three 0's and the first 1 we encounter. After that, we invert the remaining bits of the block 1010 obtaining 0101. We then obtain 01011000 as the two's complement of 10101000. Let us next look at the two's complement of the binary number 11100011. Here, the rightmost digit is 1 that we output as 1 and we invert the remaining digits, giving $(11100011)_{2's} = 00011101$. As you will see in Section 1.3.1, adding the binary numbers 11100011 and 00011101 would result in the 9-digit binary number 100000000. Since we are working in an 8-bit system, we simply ignore the leftmost bit (exactly like the digital counter would show 0000 instead of 10000) and get 00000000. Note that the above algorithm shows that the two's complement of the two's complement of N gives back N which is in line with the basic rule $-(-N) = N$.

1.2.4.4 *Back and forth between decimal and two's complement*

Conversion between decimal and two's complement representations depends on the number of bits used and the sign of the decimal value. If k is a positive integer, the two's complement representation of k is the same as its sign-magnitude binary representation. If k is a negative integer, its two's complement representation is obtained by computing the two's complement of the sign-magnitude form of $|k|$. For example, in an 8-bit machine the two's complement representation of the decimal number $k = 25$ is 00011001. For $k = -35$, the sign-magnitude representation of its absolute value is 00100011 and its two's complement is then 11011101. It is important to note that in the two's complement representation, like in sign-magnitude representation, the leftmost bit is 0 for positive numbers and 1 for negative numbers.

Converting from the two's complement representation to decimal is simple. Given an n-bit two's complement representation $b_{n-1}b_{n-2}\ldots b_1b_0$ of a decimal number k, we know that b_{n-1} represents the sign of k. If $b_{n-1} = 0$, k is positive and its decimal value is $b_{n-2}2^{n-2} + \cdots + b_12^1 + b_0$. If $b_{n-1} = 1$, k is negative and to find its decimal value, we first find its two's complement $c_{n-1}c_{n-2}\ldots c_1c_0$ which we know will correspond to the opposite $-k$ of k. We finally put a "$-$" sign in front of the answer to get the decimal value we are looking for. We clarify this with a couple of examples. Assume you are given the two 8-bit binary numbers $\alpha = 11011011$ and $\beta = 01010111$ and you are told that they are the two's complement representations of two decimal numbers A and B respectively. Your task is to find A and B. Note first that the sign digit of α is 1 indicating that A is a negative number. The two's complement of α is 00100101 which is 37 in decimal. This means that $A = -37$. As for B, we know it is positive since the sign digit of β is 0. Converting β to its decimal form gives $B = 87$.

The range of decimal numbers available for two's complement representation depends on the number of bits used. In an n-bit machine, the binary number $100 \cdots 00$ (1 followed by with $n - 1$ zeros), represents a negative number. Its two's complement remains the same. Therefore, $100 \cdots 00$ represents the decimal number -2^{n-1}. The largest positive number that can be represented in two's complement is $011 \cdots 11$ (0 followed by with $n - 1$ ones) which has a decimal value of $2^{n-2} + 2^{n-3} + \cdots + 2^1 + 2^0$. As in the proof of Proposition 1.1 above, this sum is equal to $2^{n-1} - 1$. We conclude that only decimal numbers between -2^{n-1} and $2^{n-1} - 1$ inclusively can be represented in two's complement. For example, in an 8-bit machine, this range is $[-128, 127]$.

We finish this section with showing the two main advantages of the two's complement representation:

- Since $1\underbrace{00 \cdots 00}_{n-1}$ represents -2^{n-1}, 0 is uniquely represented by $00 \cdots 00$.
- As we will see later, the two's complement representation eliminates the need for new electronic hardware to perform subtraction as it allows the use of existing hardware for addition to do subtraction.

1.3   Binary arithmetics

Standard operations on decimal numbers like addition, subtraction, multiplication and division are just particular cases of operations one can perform in any number system. Our main focus in this section is on the arithmetic of binary numbers. As explained earlier, there is more than one way to represent decimals in binary forms, so it is natural to expect that binary arithmetics will depend on the representation built inside the electronic device. Equally important is the fact that the result of any arithmetic operation is supposed to fit within the range of integers allowed by the number of bits used to store these numbers in the machine.

1.3.1 *Binary addition of unsigned integers*

Let us start by going way back (at least in my case) in our learning journey to elementary school and recall how we were first taught to add decimal numbers. Most likely, your teacher told you to start by aligning the units digits, the tens digits, hundreds digits, etc., in columns. Then, starting with the units column, add the digits in each column and if the sum is greater than 9 (that is, the sum cannot fit in one digit), generate a second digit that we *carry* to the next most significant, i.e., to the next column on the left, to be added with the digits of that column. If the sum is less than or equal to 9, the carry digit is 0. Here is a refresher.

$$\begin{array}{r} {\scriptstyle 110\ 11} \\ 925\,557 \\ +\ 76\,375 \\ \hline 1001\,932 \end{array}$$

Like in the case of decimal numbers, the sum of binary numbers involves a carry digit (of 1) when the sum of digits in a column exceeds 1. The following four sums give all rules needed to add any n-bit binary numbers: $0 + 0 = 00$, $0 + 1 = 1 + 0 = 01$, $1 + 1 = 10$ and $1 + 1 + 1 = 11$. Each addition is represented by a two-digit binary number. The digit on the right represents the "output" of the addition called the *sum digit* and the digit on the left is the carry digit that we add to the digits of the column on the left. Here is an example of adding two binary numbers, where the digits in smaller font in the top row are, as in the decimal case, the carry digits.

$$\begin{array}{r} {\scriptstyle 1111\ 111} \\ 1100\,1001 \\ +\ 1111\,1111 \\ \hline 11100\,1000 \end{array}$$

Unlike the "pencil and paper" addition shown above, machines have to work within certain limits and it could happen (like in the above example) that the result of the addition exceeds the number of storage units allowed. This is a situation known as *overflow*. Detecting overflow in unsigned addition is simple: a carry out of 1 from adding the last significant bits indicates that an overflow has occurred. Take for instance the (unsigned) addition $0111 + 1001$ (corresponding to $7 + 9$ in decimal) in a 4-bit machine, which results in 10000. As the carry out is 1 from the leftmost bit, an overflow has occurred. In fact, $7 + 9 = 16$ exceeds the maximum (15) allowed by a 4-bit machine.

1.3.2 *Binary addition of signed integers*

Modern machines use the two's complement scheme to represent signed numbers. Two's complement representation allows binary arithmetic to be performed without worrying much about the signs of the operands. In this section, we focus on binary addition using the two's complement format.

Let A and B be two integers (in decimal) and let $(A)_{2's}$ and $(B)_{2's}$ be their respective representations in two's complement. To perform $A+B$, the machine starts by computing the unsigned addition of $(A)_{2's}$ and $(B)_{2's}$. That is, it treats $(A)_{2's}$ and $(B)_{2's}$ as unsigned numbers including their sign bits. Any carry out bit from the addition in the leftmost column is then ignored. Let us look at some examples using an 8-bit machine. Assume $A = 52$, $B = -37$. To find $A+B$, we write A and B in two's complements: $(A)_{2's} = 00110100$ and $(B)_{2's} = 11011011$. We then perform the sum $(A)_{2's} + (B)_{2's}$ as unsigned binaries:

$$
\begin{array}{r}
{}^{111} \\
0011\,0100 \\
+\ 1101\,1011 \\
\hline
1\,0000\,1111
\end{array}
$$

Since the result has 9 bits, one more than the storage limit, we simply drop the leftmost bit and get 00001111, which is 15 in decimal. The answer is correct: $52 + (-37) = 15$. Let us now consider an example of adding two negative numbers: Assume $A = -15$, $B = -24$, then $(A)_{2's} = 11110001$ and $(B)_{2's} = 11101000$. Adding $(A)_{2's}$ and $(B)_{2's}$ (as unsigned) gives 11011001 (after dropping the carry out 1 from the leftmost bit). Note that $(11011001)_{2's} = -39$, which is the correct answer: $-15 + (-24) = -39$.

Unlike the unsigned addition, detecting an overflow in two's complement addition requires a bit more attention. First note that an overflow can *never* occur when the two operands are of opposite signs. The reason is that the sum in this case is smaller than one of the operand and since both operands fit within the available range of numbers, the sum must fit in that range as well. If A and B have the same sign (both positive or both negative), then an overflow can occur if the result of the addition has the opposite sign of that of the operands (in this case, the answer is incorrect). In other words, to detect an overflow in two's complement addition, examine the sign bit of the result: if it is different from that of the operands, an overflow has occurred and if not there is no overflow. Note that this is equivalent to say that an overflow occurs only if the carry bit into the leftmost column is not the same as the carry out bit from that column. This last observation allows an easy overflow detection hardware design in machines. Let us next consider some examples of addition using 8-bit two's complement:

$$
\begin{array}{cccc}
\overset{1}{} & \overset{1\ \ 1\ 11}{} & \overset{111}{} & \overset{1\ \ 1\ 1}{} \\
0100\,0100 & 1010\,1011 & 0011\,0100 & 0010\,1100 \\
+\,0110\,0000 & +\,1010\,0110 & +\,1101\,1011 & +\,0010\,1101 \\
\hline
1010\,0100 & 10101\,0001 & 10000\,1111 & 0101\,1001
\end{array}
$$

The first sum has an overflow without carry out. It gives a negative answer to the sum of two positive numbers (the sign bit of the answer is 1 while both the operands have 0 as leftmost bit). The second sum presents also an overflow but with a carry out (by dropping the leftmost bit, the sum gives a positive answer while the operands are both negative). A carry out with no overflow occurs in the third addition since the operands are of opposite signs (we simply drop the leftmost bit of the answer). There is no carry out nor an overflow in the last sum.

1.3.3 *Two's complement subtraction*

The reason why two's complement representation is so popular in computer designs is because when signed numbers are added or subtracted in this format, they can be treated as unsigned numbers with any carry out from the last bit dropped. The answer is correct regardless of the signs of the operands as long as it fits within the number of bits allowed. Unlike addition, subtraction is not a commutative operation. That is $A - B$ and $B - A$ result in two different values. In the subtraction $D = A - B$, A is called the *minuend*, B the *subtrahend* and D is called the *difference*. To perform the subtraction $A - B$, write the two's complement form of the subtrahend and then add the answer to the two's complement form of the minuend using the relation $A - B = A + (-B)$. Any addition carry out from the sign bit is simply ignored. This gives the two's complement representation the advantage of performing both addition and subtraction using the same *hardware* as we will see later. For example, to perform the subtraction $55 - 78$ in two's complement with 8 bits, we start by writing both 55 and -78 in two's complement: $(55)_{2's} = 00110111$, $(-78)_{2's} = 10110010$. We then perform the addition:

$$
\begin{array}{c}
\overset{11\ \ 11}{} \\
0011\,0111 \\
+\,1011\,0010 \\
\hline
1110\,1001
\end{array}
$$

There is no carry out nor an overflow in this case. The result is a negative number since its sign bit is 1 and its decimal value is -23 (obtained by converting 11101001 from two's complement to decimal as we did in Section 1.2.4.4) which is the correct answer: $55 - 78 = -23$.

1.4 Logic

In the context of this chapter, Logic is the set of mathematical rules governing any electrical circuit design for binary arithmetic in a machine. You will be amazed to learn that basic words like "AND", "OR", "NOT" can be interpreted as "switches" and "gates" inside your computer.

Any human language is usually a collection of words and symbols that one can put together using a set of grammatical rules to form different types of sentences. In the English language for example, we have the imperative sentence, the interrogative sentence and the declarative sentence among others. What distinguishes declarative sentences from the others is the fact that they carry a *truth value*. Every declarative sentence is either True (T) or False (F). From the Logic point of view, a declarative sentence is called a *proposition*. For instance, the sentence "Two to the power of four is equal to 18" is a proposition with truth value F and "The number of binary strings with 10 bits is 2^{10}" is a proposition with truth value T. The ease of determining the truth value of the previous two propositions is just an illusion and things can get complicated very quickly as we combine propositions together to form new ones. Consider for example the following proposition: "The hunt is over or the lion is dead if and only if it rains in the forest and the moon is full unless the hunter hides behind the trees". In this latest complex proposition, words like "or", "if and only if", "and", "unless" are called *logic connectives* or *logic operators*. Their role is to connect propositions together to make new compound ones. Table 1.2 gives the names and symbols of the basic logic connectives. Except for the connective "NOT", each of these connectives requires two input propositions that we denoted as p and q in the table.

When we combine propositions using the above operators, the result remains a proposition and as such it is either True or False. Without going into too much details, we explain briefly the output value (T or F) of some of these operators. The compound proposition $p \vee q$ (p OR q) is always true except in the case where p and q are both false. On the other extreme,

Table 1.2   Basic logic connectives.

Connective Name	Symbol	English Expressions
Disjonction (OR)	$p \vee q$	p or q, p unless q
Conjonction (AND)	$p \wedge q$	p and q, p but q
Implication	$p \rightarrow q$	p implies q
Biconditional	$p \leftrightarrow q$	p if and only if q
Exclusive OR (XOR)	$p \oplus q$	p or q but not p and p
Negation (NOT)	$\neg p$	Not p

Table 1.3   A truth table.

p	q	$p \vee q$	$p \wedge q$	$p \rightarrow q$	$p \leftrightarrow q$	$p \oplus q$	$\neg p$	$\neg q$
T	T	T	T	T	T	F	F	F
T	F	T	F	F	F	T	F	T
F	T	T	F	T	F	T	T	F
F	F	F	F	T	T	F	T	T

$p \wedge q$ (p AND q) is always false except in the case where p and q are both true. To explain the truth value of the "XOR" operator ($\oplus$), imagine I say "Joseph is teaching a class or he is gone fishing", then I would be saying the truth if *exactly one* of the components "Joseph is teaching a class", "he is gone fishing" is true. Clearly the statement is false otherwise. Table 1.3 gives the truth values of the above logic operators as functions of the truth values of their components.

So what does this "linguistic" introduction have to do with electronic and circuit design? Like a proposition, each bit can take two values 0 or 1 and each switch in a digital circuit is also under two possible states: high voltage or low voltage. Think of Logic as being the "brain" of any electronic circuit that sends signals to different parts of the circuit to execute various tasks.

1.4.1   *Logic gates*

The OR, AND, XOR, and NOT operators described above can be thought of as electronic "gates" in a circuit producing a certain output signal based on the status of their binary input signals. Logic gates are the building blocks of any electronic design since they have the power to allow or block a binary signal through parts of the circuit. As one engineer once explained them to me, they are the "decision makers" in electronic circuits. Industries use graphic symbols to represent various logic gates. The following gives the most common graphic representations of OR, AND, XOR, and NOT gates.

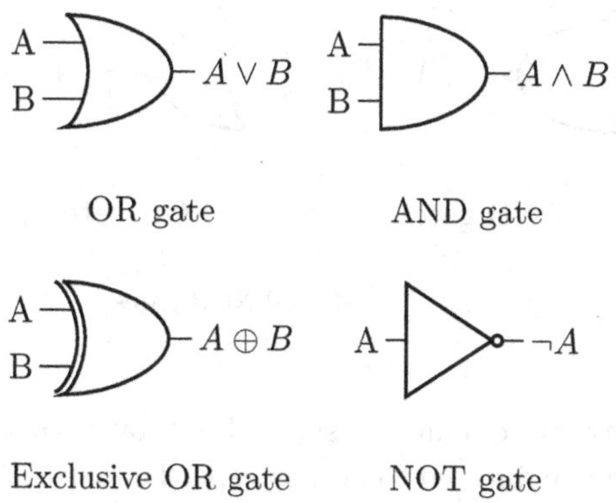

A —⟩ $A \vee B$ A —⟩ $A \wedge B$
B —⟩   B —⟩

OR gate   AND gate

A —⟩ $A \oplus B$ A —▷o— $\neg A$
B —⟩

Exclusive OR gate   NOT gate

Fig. 1.1   Basic logic gates.

If you look at a map of a digital electronic chip, you will most likely see more logic gates symbols than the four listed in Figure 1.1 above. For instance in TTL technology (Transistor-Transistor Logic), the NAND gate plays a crucial role. The NAND gate can be interpreted as an AND gate followed by a NOT gate. If A, B are binary inputs, then the output of a NAND gate is $\neg(A \wedge B)$:

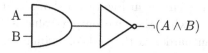

Fig. 1.2   NAND gate action.

In CMOS technology (Complementary Metal-Oxide-Semiconductor), an important logic gate is the NOR gate. Similar to the NAND gate, the NOR gate is an OR gate followed by a NOT gate. If A, B are binary inputs, then the output of a NOR gate is $\neg(A \vee B)$:

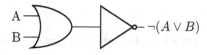

Fig. 1.3   NOR gate action.

The NAND and NOR gates are represented graphically as follows.

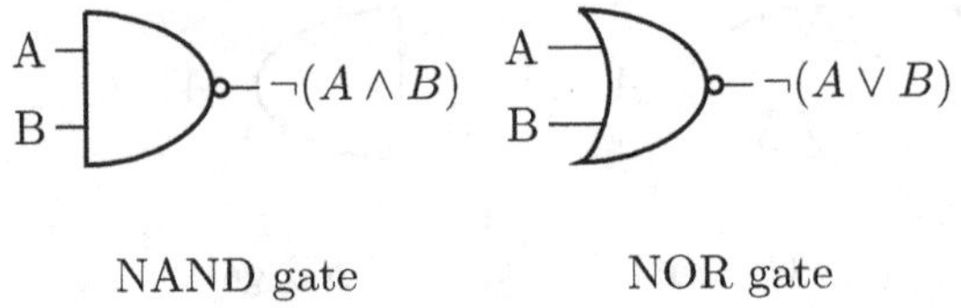

NAND gate   NOR gate

Fig. 1.4   NAND and NOR gates.

A typical electronic circuit is usually built by a complex network of logic gates carefully engineered to perform specific tasks. Depending on the complexity of the circuit, logic gates are usually built with more than two binary input terminals in order to accommodate more complex operations through the gate. Figure 1.5 shows an example of a logic circuit with four binary inputs x_0, x_1, x_2 and x_3 and with logic gates having multiple input entries. The network is designed to produce a binary output function f in response to various combinations of binary inputs. The filled little circles in the network ($\bullet$) indicate that the wires are actually connected and electrons can flow in both wires to the corresponding gates. Unconnected wires are simply indicated by intersecting lines ($\perp$) with no filled circle at the intersection point.

A quick look at the circuit in Figure 1.5 shows that it could be cumbersome to determine what the final output value is for a given list of binary inputs. Even harder is the task of finding an algebraic expression $f(x_0, x_1, x_2, x_3)$ for the output as a function of the four inputs. The other side of this problem is of more importance from a design point of view: in many instances (like an alarm system, traffic lights, binary adder, ...) we can actually come up with the desired output as a function $f(x_0, x_1, x_2, \ldots, x_n)$ of the input variables but the challenge is in building a circuit to implement it. Another question to answer is, "How to ensure that the design is the most cost effective in terms of number of gates and devices used considering the fact that a given binary output can be achieved using several designs?"

Many tools were developed through the years to come up with the best circuit design to perform a given task. Some of these tools are graphical, like the Karnaugh map and the semantic tableaux and others are algebraic like the Boolean Algebra axioms, the sum of products and the product of sums. In this chapter, we focus on the algebraic tools as they are more representative of the power of (what looks like) abstract mathematics in circuit designs.

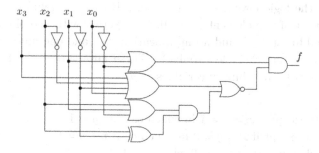

Fig. 1.5   A logic circuit with four binary inputs.

1.5   Boolean Algebra

Table 1.4   Properties of the operations of a Boolean Algebra.

Property	Name
$x + y = y + x$	Commutativity of OR
$xy = yx$	Commutativity of AND
$x + (y + z) = (x + y) + z$	Associativity of OR
$x(yz) = (xy)z$	Associativity of AND
$x + 0 = x$	OR identity element
$x1 = x$	AND identity element
$x + 1 = 1$	Output set
$x0 = 0$	Output reset
$x + (yz) = (x + y)(x + z)$	OR distributive law
$x(y + z) = (xy) + (xz)$	AND distributive law
$(x + y)' = x'y'$	De Morgan's NOR law
$(xy)' = x' + y'$	De Morgan's NAND law
$x + x = x$	OR idempotent law
$xx = x$	AND idempotent law
$x + x' = 1$	Complementation rule for addition
$xx' = 0$	Complementation rule for multiplication
$(x')' = x$	Double negation

In the mid 1800's, the British mathematician George Boole came up with a mathematical system to model the logic laws in an algebraic way. After almost a century of refinements and improvements on the system by various mathematicians and algorithm designers, the scheme finally found its way into real world applications and became an important tool in engineering. In our days, people refer to Boole's system as the *Boolean Algebra*. Inspired by the logic laws described above, a Boolean Algebra can be described as a set of two elements, say $\{0, 1\}$, together with two binary laws written as addition " $+$ " and a multiplication " $\cdot$ " and one unary law "$'$" (operation requiring a single variable). These laws satisfy the following rules or *axioms* for any binary variables x, y:

(1) $x \cdot y = 0$ except if $x = y = 1$ in which case $x \cdot y = 1$
(2) $x + y = 1$ except if $x = y = 0$ in which case $x + y = 0$
(3) $x' = 1$ when $x = 0$ and $x' = 0$ when $x = 1$.

At this point, chances are you started to draw a link between the operations of a Boolean Algebra and the logic operators defined above. In fact, in a Boolean Algebra, the expression $x + y$ is read "x or y", $x \cdot y$ is read "x and y" and "x'" is read "not x". From this point of view, the above axioms become more natural. As in the case of real numbers, the multiplication will be denoted from this point on by juxtaposition of operands, so $x \cdot y$ is replaced by xy. Also similar to usual arithmetics, there are some precedence rules for the order the Boolean operations. In the order of the highest to the lowest precedence, the rules are: the parentheses, the NOT operation ("$'$"), the AND operation (the multiplication), the OR operation (the addition). An expression combining binary variables with one or more of the above laws is often referred to as a *function*. For instance $f(x, y, z) = x + x'yz + y'$ is a function that takes the value 1 for the input values $x = y = 0$ and $z = 1$. In addition to the above axioms, the operations of a Boolean Algebra satisfy the important properties found in Table 1.4 on page 19.

Many of these laws may seem very obvious or even trivial to you at this point. But remember that their main use in this context is in simplifying the output function of a complex circuit and from this point of view they could be sometimes tricky to be handled efficiently. In what follows, we work out an example to show how Boolean Algebra is used to simplify a certain digital circuit. Figure 1.6 shows a logic circuit with three binary inputs x, y and z.

Following the first AND gate from the top, it is easy to see that its input is $x' \wedge y' \wedge z$ or $x'y'z$ in Boolean notations. Similarly, the inputs for the other AND gates (starting from the second one on top) are $x'yz'$, $xy'z$, xyz and xyz' respectively. All five gates are connected to an OR gate making the final circuit output $x'y'z + x'yz' + xy'z + xyz + xyz'$. We obviously wish to come up with a simpler circuit that gives exactly the same output for every list of binary values for x, y and z. The Boolean Algebra properties become very handy for this task. The details of the calculations in Figure 1.7 are not hard to follow and left to the reader to verify. The last expression $y'z + yz' + xy$ shows that we can replace the circuit in Figure 1.6 with the following equivalent circuit but more efficient in terms of the number of logic gates.

1.5.1 *Sum of products - Product of sums*

We start with some terminology. Given a set $S = \{x_1, \ldots, x_n\}$ of logic variables, we define a *minterm* over S as being a product of the form $y_1 y_2 \cdots y_n$ where each y_i is either equal to x_i or the complement of x_i (that is x_i'). For each i, the variable x_i appears *exactly once* in each minterm either as x_i or as its complement x_i' but not both. As a consequence, there is a total of 2^n minterms over S. For example, the following are all the $2^3 = 8$ minterms over the variables x, y and z:

$$x'y'z', \; x'y'z, \; x'yz', \; x'yz$$
$$xy'z', \; xy'z, \; xyz', \; xyz.$$

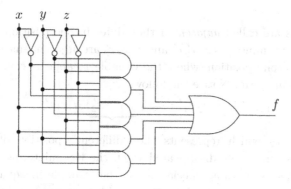

Fig. 1.6   A logic circuit with three binary inputs.

$$x'y'z + x'yz' + xy'z + xyz + xyz'$$
$$= (x'y'z + xy'z) + (x'yz' + xyz') + xyz$$
$$= (x' + x)y'z + (x' + x)yz' + xyz$$
$$= y'z + yz' + xyz$$
$$= y'z + y(z' + xz)$$
$$= y'z + y[(z' + x)(z' + z)] \quad \text{(OR distributive law)}$$
$$= y'z + y(z' + x)$$
$$= y'z + yz' + xy.$$

Fig. 1.7   Calculations in a Boolean Algebra.

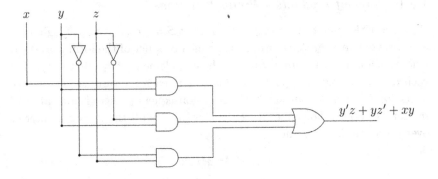

Fig. 1.8   A logic circuit equivalent to the circuit in Figure 1.6 with fewer gates.

Two minterns are called *adjacent* if they differ in one position only. For instance, the two minterms $x_1x_2'x_3'$ and $x_1x_2x_3'$ are adjacent since they only differ at the second position, where the variable x_2 appears complemented in the second minterm. Notice the following:

$$x_1x_2x_3' + x_1x_2'x_3' = x_1x_3'\underbrace{(x_2 + x_2')}_{=1} = x_1x_3'. \quad (1.1)$$

The variable x_2 which represents the "different" position of the two minterms has completely disappeared when the two minterms are added together. There is nothing special about the example in equation (1.1) and every time two adjacent minterms are added, we can simplify the sum by dropping the different position. Exploiting this property of adjacent minterms will play a crucial role in simplifying algebraic expressions (it is at the heart of a schematic method to simply Boolean expressions known as Karnough maps that we will not explore in this book). A minterm is True for exactly one combination of variable inputs. For example, the minterm $x_1'x_2'x_3$ is only true for $x_1 = 0$, $x_2 = 0$ and $x_3 = 1$.

The dual notion of a minterm is the *maxterm*. A maxterm over the set $S = \{x_1, ..., x_n\}$ of logic variables is a sum of the form $y_1 + y_2 + \cdots + y_n$ where each y_i is either x_i or x_i'. Like the minterms, there are 2^n maxterms over n logic variables. A maxterm is False for exactly one combination of variable inputs.

1.5.2 Sum of products

The Boolean expression

$$f = x'y'z + x'yz' + xy'z + xyz + xyz' \tag{1.2}$$

representing the output for the circuit in Figure 1.6 above is called a *sum of products form* of f. The name is self explanatory. Note that in this expression of f, each of the products $x'y'z$, $x'yz'$, $xy'z$, xyz, and xyz' is a minterm but that is not necessarily true in every sum of products form. For instance, the above calculations that led to the diagram in Figure 1.8 show that an equivalent form of the same function f is given by $y'z + yz' + xy$ which we still call a sum of products form of f. In other words, there are several ways to write a Boolean expression as a sum of products and a sum of minterms is one of them. There is however a unique way to express a Boolean expression as a sum of minterms. Writing the (unique) sum of minterms form of a Boolean expression f from the truth table of f can be achieved by writing the minterm corresponding to each row in the table where the output of f is 1 and then adding all the minterms. Although this method provides an easy way to derive a sum of products expression, it is not optimal in the sense that it does not produce the simplest expression for f (as we saw in Figures 1.6 and 1.8) but it is a starting point and Boolean Algebra manipulations can then be used to simplify it. Let us look at an example. Suppose you are given the truth table of a Boolean function $f(x, y, z)$, see Table 1.5.

The truth value of f is 1 in the first, fifth and eighth rows of the table. The minterms corresponding to these rows are $x'y'z'$, xyz' and $xy'z'$ respectively. The sum of minterms form of f is then given by

Table 1.5 A Boolean function.

x	y	z	f
0	0	0	1
0	0	1	0
0	1	1	0
0	1	0	0
1	1	0	1
1	1	1	0
1	0	1	0
1	0	0	1

$f(x, y, z) = x'y'z' + xyz' + xy'z'$. Now, Boolean Algebra properties can be used to simplify this expression:

$$x'y'z' + xyz' + xy'z' = (x'y'z' + xy'z') + xyz'$$
$$= \underbrace{(x' + x)}_{1} y'z' + xyz'$$
$$= y'z' + xyz'.$$

A useful sum of products we will use in the adder circuit design below is that of $x \oplus y$, i.e. the exclusive OR. From the truth table of basic logic operators given above, it is easy to see that the sum of products of $x \oplus y$ is $x'y + xy'$. Note also that $(x \oplus y)' = (x'y + xy')' = xy + x'y'$ and here is why:

$$(x \oplus y)' = (x'y + xy')' = (x'y)'(xy')' \text{ (by DeMorgan laws)}$$
$$= (x + y')(x' + y) \text{ (by DeMorgan laws)}$$
$$= \underbrace{xx'}_{=0} + xy + y'x' + \underbrace{yy'}_{=0}$$
$$= xy + x'y'.$$

1.5.3 *Product of sums*

We start by noting first that minterms and maxterns are actually not strangers to each others. One is the complement of the other. The maxterm $x + y' + z'$ is the complement of the minterm $x'yz$ since $(x'yz)' = x + y' + z'$ by the DeMorgan's Laws. From this perspective, every Boolean expression can also be written as the product of maxterms in a unique fashion. Product of maxterms is a special type of a *product of sums*. From the truth table, the product of maxterms form can be found as follows: first select all rows in the table where the function output is 0. For each of these rows, form the associated maxterm by complementing the minterm that corresponds to that row. For example, if $x = 1$, $y = 1$, $z = 0$ is one row in the truth table where the output of f is 0, then the corresponding maxterm is $(xyz')' = x' + y' + z$. The product of maxterms form of f is then obtained by multiplying all the corresponding maxterms. Let us look again at the truth table of the function f in the previous section. The output of f is 0 in rows $2, 3, 4, 6$ and 7. The maxterm corresponding to row 2 is $(x'y'z)' = x + y + z'$. Similarly, the maxterms corresponding to rows $3, 4, 6$ and 7 are $(x+y'+z')$, $(x+y'+z)$, $(x'+y'+z')$ and $(x'+y+z')$ respectively. The product of maxterms form of f is then

$$(x + y + z')(x + y' + z')(x + y' + z)(x' + y' + z')(x' + y + z').$$

Both the sum of minterms and the product of maxterms are called *standard forms* of the Boolean expression.

1.6 Digital adders

We now arrive at the heart and soul of this chapter. What is the procedure that an electronic calculator follows in order to perform a certain arithmetic operation? The accurate answer to this question requires a fair knowledge of all components of an electronic circuit and the physical laws that enable these components to work effectively together. Of course, this is well beyond the scope of this book. Components such as the processor inside a machine are physical implementation of mathematical ideas and designs. Our main interest is to look beyond the hardware in order to understand the mathematical ideas.

1.6.1 *Half-adder*

As we saw in Section 1.3.1, when two binary digits x, y are added, a sum $\sum$ and a carry C are produced. If you think of $\sum(x, y)$ and $C(x, y)$ as Boolean expressions of the variables x and y, then the basic rules of adding bits seen in Section 1.3.1 give their truth tables as illustrated in Table 1.6.

Note that the carry is always 0 except in the case where both digits are equal to 1. This is exactly the output of the AND operator. So,

Table 1.6   Sum and carry.

x	y	Sum ($\sum$)	Carry (C)
0	0	0	0
1	0	1	0
0	1	1	0
1	1	0	1

$C(x, y) = x \wedge y$ or simply xy using the Boolean notation. As for the sum $\sum$, note that its outcome is 1 if the value of *exactly* one of the variables is 1 and the output is 0 otherwise. This is precisely the output of the Exclusive OR operator. Therefore $\sum(x, y) = x \oplus y$. In light of these facts, we can construct a logic circuit to implement this addition of two bits using one AND gate and one Exclusive OR gate. The circuit takes in two inputs (binary bits) and gives out two outputs, a sum $\sum$ and a carry out C:

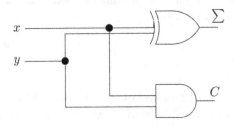

Fig. 1.9   Half-adder circuit.

1.6.2   *Full-adder*

While the half-adder is probably the simplest adder circuit, it has a major handicap. It has only two inputs which means it cannot deal with the carry in that usually occurs in binary addition. This makes the half-adder capable of performing binary addition only when there is no carry out (a carry out of 0), hence the name. The building block of any digital adder requires a system that takes into account the two bits and the carry from the previous column. This is what a *full-adder* is designed to do. In the full-adder design, we have three inputs: the two bits to be added and the carry in from previous addition. As in the case of the half-adder, there are two outputs: a sum and a carry out. To distinguish between the carry in and the carry out in the fuller-adder, we write C_{in} for the carry in and C_{out} for the carry out.

Table 1.7 The full-adder.

x	y	C_{in}	Sum ($\sum$)	(C_{out})
0	0	0	0	0
0	0	1	1	0
0	1	1	0	1
0	1	0	1	0
1	1	0	0	1
1	1	1	1	1
1	0	1	0	1
1	0	0	1	0

The truth table of the full-adder is found in Table 1.7. Note that $1 + 1 + 1 = 11$ (sum of 1 and a carry of 1).

From the truth table, we can construct Boolean expressions for the sum $\sum$ and the carry out C_{out}. The idea is to write first the *sum of products* for $\sum$ and C_{out} and then make some simplifications using Boolean Algebra properties.

$$\sum = x'y'C_{in} + x'yC'_{in} + xyC_{in} + xy'C'_{in}$$

$$C_{out} = x'yC_{in} + xyC'_{in} + xyC_{in} + xy'C_{in}.$$

Using the relations established in Section 1.5.2, we can simplify $\sum$ and C_{out} as follows:

$$\begin{aligned}
\sum &= x'y'C_{in} + x'yC'_{in} + xyC_{in} + xy'C'_{in} \\
&= (x'y' + xy)C_{in} + (x'y + xy')C'_{in} \\
&= (x \oplus y)'C_{in} + (x \oplus y)C'_{in} \\
&= (x \oplus y) \oplus C_{in}
\end{aligned}$$

and

$$\begin{aligned}
C_{out} &= x'yC_{in} + xyC'_{in} + xyC_{in} + xy'C_{in} \\
&= xy \underbrace{(C'_{in} + C_{in})}_{=1} + (x'y + xy')C_{in} \\
&= xy + (x \oplus y)C_{in}.
\end{aligned}$$

The circuit in Figure 1.10 implements the above equations using XOR and AND gates.

Full-adders are the building blocks in any electronic device capable of doing binary arithmetics. Recall that binary addition of two n-bit numbers

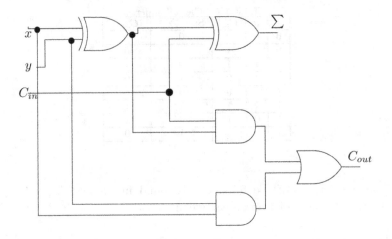

Fig. 1.10   Full-adder circuit.

is done by adding first the two least significant bits and progressing to the addition of the most significant bits. In the process, the carry out produced at any stage is added to the two bits in the next position. This can be designed in the machine by cascading together n full-adders, one adder for each pair of bits. The carry out produced by the sum of each pair of bits "ripples" through the chain until we get to the last carry out. Such an adder is called *carry ripple adder*. Figure 1.11 shows a *block design* for a carry ripple adder for 3-bit binary numbers $A_2A_1A_0$ and $B_2B_1B_0$. Each rectangle contains a full-adder design as shown in Figure 1.10. Note that if the first carry-in (C_0) is 0, then it is represented by a "ground" in real designs.

The diagram in Figure 1.12 is a detailed logic implementation of a 3-bit ripple carry adder with XOR, OR and AND gates. The circuit adds the two binary numbers $A_2A_1A_0$ and $B_2B_1B_0$ and outputs the binary number $\Sigma_3\Sigma_2\Sigma_1\Sigma_0$.

1.6.3   Lookahead adder

In a carry ripple adder, all bits are entered at the same time in the circuit (as seen in Figure 1.11 below) and the sum of bits at position i cannot be executed until all carries have rippled through the previous positions.

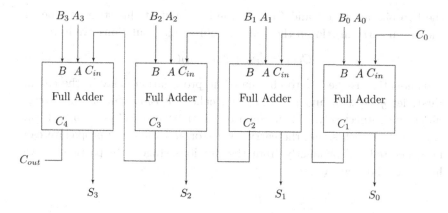

Fig. 1.11   3-bit carry ripple adder.

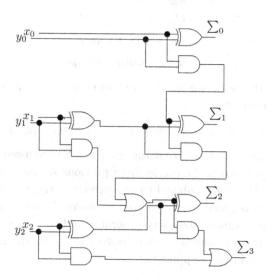

Fig. 1.12   A 3-bit ripple adder logic circuit.

Considering the significantly large number of bits computers are expected to deal with, the accumulated wait times can cause a serious delay. One solution to speed up the process is to try to generate the carry-input locally at each stage to avoid waiting for its value to ripple through. Let A_i, B_i be the two bits at stage i and C_{i+1} be the carry-out at this stage (C_0 being the initial carry in). From the above algebraic manipulations, we have

$$C_{i+1} = x_i y_i + (x_i \oplus y_i)C_i. \qquad (1.3)$$

This equation is the key to eliminate the propagation delay in the chain. First, let $g_i = x_i y_i$ and $p_i = x_i \oplus y_i$, called *carry-generator* and *carry-propagate* respectively. Then equation (1.3) becomes $C_{i+1} = g_i + C_i p_i$. Note that at each stage, the carry-generator and the carry-propagate can be generated independently from the previous stages. In particular, we have the following values:

$$\begin{aligned}
C_1 &= g_0 + C_0 p_0 \\
C_2 &= g_1 + C_1 p_1 \\
&= g_1 + (g_0 + C_0 p_0)p_1 \\
&= g_1 + g_0 p_1 + C_0 p_0 p_1 \\
C_3 &= g_2 + C_2 p_2 \\
&= g_2 + (g_1 + g_0 p_1 + C_0 p_0 p_1)p_2 \\
&= g_2 + g_1 p_2 + g_0 p_1 p_2 + C_0 p_0 p_1 p_2.
\end{aligned}$$

Continuing this way, we get the following general expression for the carry out at the ith stage:

$$C_{i+1} = g_i + g_{i-1}p_i + g_{i-2}p_i p_{i-1} + \cdots + C_0 p_i p_{i-1} p_{i-2} \ldots p_0. \qquad (1.4)$$

This new expression shows that the carry can be produced at any stage without the need to know the carries from previous stages and hence avoid the time delay. An adder designed to produce the carry digit locally based on equation (1.4) is called a *carry lookahead adder*. In spite of the calculations involved to compute the carry-generator and the carry-propagate at each stage, the carry lookahead adder design remains much faster than the ripple carry adder in large applications.

1.6.4 *Two's complement implementation*

In Section 1.2.4.3, an algorithm to find the two's complement of a binary number was introduced. Given an n-bit binary number $N = N_{n-1} \ldots N_0$, the algorithm requires to check the values of first bits starting at N_0 until we reach the first 1. That could be a challenge to implement in a machine. In reality, computers use a different approach to compute the two's complements. Recall that the two's complement $N_{2's}$ of N is obtained by finding first the one's complement $N_{1's}$ of N and then adding 1 to the result: $N_{2's} = N_{1's} + 1$ and $N_{1's}$ is obtained from N by just reversing its bits. From a technical point of view, both operations of reversing the bits and adding 1 are easy to implement in a machine.

1.6.5 *Adder-subtractor combo*

As mentioned in Section 1.2.4.3, one advantage of representing signed integers in two's complement format is in the flexibility to use an existing adder circuit to perform subtraction. This will certainly reduce the number of gates used and eliminate some of the complexity of wiring the circuit. In this section we explain in a bit more detail how an adder-subtractor combo circuit can be constructed and the mathematics behind this design. Figure 1.13 shows a design for an adder/subtractor circuit for two 4-bit binary numbers $A = A_3 A_2 A_1 A_0$ and $B = B_3 B_2 B_1 B_0$. In the design, S is a switch that controls the signal flow in the circuit and hence allows the transition between the addition and subtraction modes as we will see below. Each of the rectangles labeled "FA" represents a full-adder.

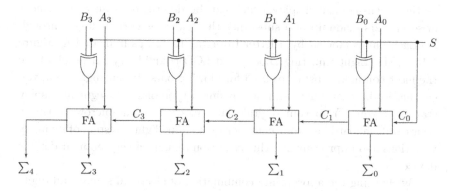

Fig. 1.13   4-bit adder/subtractor combo.

If the switch S is turned off (that is the value of S is 0 or low voltage through S), then the original input C_0 is 0 and the ith XOR gate ($i = 0, 1, 2, 3$) outputs the value $B_i \oplus 0$. Note that

$$B_i \oplus 0 = (B_i \vee 0) \wedge \neg (B_i \wedge 0)$$
$$= B_i \wedge \neg 0$$
$$= B_i \wedge 1 = B_i.$$

This means that the XOR gates output $B = B_3 B_2 B_1 B_0$ and each output $\sum_i$ is $A_i + B_i$. The circuit acts like an adder in this case and performs the operation $A + B$. If the value of S is 1 (high voltage through S), then the ith XOR gate outputs $B_i \oplus 1$. Now,

$$B_i \oplus 1 = (B_i \vee 1) \wedge \neg(B_i \wedge 1)$$
$$= 1 \wedge \neg B_i = \neg B_i = B_i'.$$

The XOR gates are then just bit invertors producing the one's complement B' of B. Since the original input C_0 is 1, the circuit performs the operation $A + B' + 1$ which amounts to adding A to the two's complement of B (Section 1.6.4). In other words, the circuit performs $A - B$ and it is a subtractor in this case.

1.7 BCD to seven-segment decoder

After exploring some of the techniques used by your calculator to perform arithmetical operations, it is time to look at how the calculator actually displays the results on its screen. An early device to achieve this is known as the seven-segment display which can be described as an arrangement of seven bars made up of a substance that glows as current flows through it. Each bar is labeled by a letter from "a" to "g" as indicated in Figure 1.14. LED (Light-Emitting Diode) and LCD (Liquid Crystal Display) are the most common substances used for glowing bars. In our digital era, you can hardly be somewhere and not having a type of seven-segment display staring at you. From your digital alarm clock, to your microwave timer, your car dashboard and your neighborhood traffic light. In spite of its many variations and improvements, this invention remains a very popular display device.

By choosing the appropriate combination of electrical signals, all digits of the decimal system (0 to 9) can be produced as shown in Figure 1.15 by turning on the corresponding LED bars.

Along with the 10 digits, an arrangement of several seven-segment displays can be used to represent multi digits integers and other characters like letters, commas, dots and others. The idea is to turn on some parts to produce the desired character while keeping the other parts off. BCD to a seven-segment converter is an early form of display, more recent displays use a multitude of pixels to create more sharp looking characters. In this section, we focus on the circuit design for a single classic seven-segment

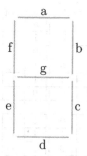

Fig. 1.14 Seven-segment LED display.

Fig. 1.15 All 10 digits on a seven-segment display.

LED display.

In a seven-segment display design, the 10 digits 0 to 9 are represented using their BCD codes (see Section 1.2.3 above). The circuit, known as *BCD to seven segment converter*, must then convert a 4-bit input $x_3x_3x_1x_0$ into a 7-bit output that can be used to turn on specific bars while keeping others unlit. Each of the bars "a" through "g" on a seven-segment display can be thought of as a logic function of 4 variables x_0, x_1, x_2 and x_3. For instance, digit 3 is represented by the BCD code 0011 and in order to display it on the screen, bars "a", "b", "c", "d" and "g" must be turned on (binary value 1, electrical signal passing through) and bars "f" and "e" turned off (binary value 0, no electrical signal through). So our circuit must convert the BCD code 0011 into 1111001. Using the same "visual inspection" for the ten digits and their representations on the seven segment LED display, it is straightforward to verify the truth tables of each of the seven bars given in Table 1.8 on page 34.

The next task is to come up with a simple Boolean expression for each of the logic functions "a" through "g". We will work out the details for bar "a" leaving the expressions for the other bars as exercises for the reader. From Table 1.8, we can see that it is easier to write the standard product of sums expression for "a" rather than the sum of product since the output of

Table 1.8   Truth tables of each of the seven bars.

Digit	x_0	x_1	x_2	x_3	a	b	c	d	e	f	g
0	0	0	0	0	1	1	1	1	1	1	0
1	0	0	0	1	0	1	1	0	0	0	0
2	0	0	1	0	1	1	0	1	1	0	1
3	0	0	1	1	1	1	1	1	0	0	1
4	0	1	0	0	0	1	1	0	0	1	1
5	0	1	0	1	1	0	1	1	0	1	1
6	0	1	1	0	1	0	1	1	1	1	1
7	0	1	1	1	1	1	1	0	0	0	0
8	1	0	0	0	1	1	1	1	1	1	1
9	1	0	0	1	1	1	1	1	0	1	1

"a" is 0 at only two inputs combinations (see Section 1.5.1). The product of maxters of "a" is given by

$$a = (x_0 + x_1 + x_2 + x_3')(x_0 + x_1' + x_2 + x_3). \tag{1.5}$$

We use the Boolean Algebra properties to come up with a simplified version of expression (1.5):

$$a = (x_0 + x_1 + x_2 + x_3')(x_0 + x_1' + x_2 + x_3)$$

$$= \underbrace{x_0 x_0}_{=x_0} + x_0 x_1' + x_0 x_2 + x_0 x_3 + x_1 x_0 + \underbrace{x_1 x_1'}_{=0} + x_1 x_2 + x_1 x_3$$

$$+ \underbrace{x_2 x_0 + x_2 x_1' + x_2 x_2}_{=x_2} + x_2 x_3 + x_3' x_0 + x_3' x_1' + x_3' x_2 + \underbrace{x_3' x_3}_{=0}.$$

Next, we regroup terms using associativity and commutativity of both multiplication and addition starting with the single terms x_0 and x_2:

$$a = x_0 + x_2 + x_1 x_3 + x_1' x_3' + (x_0 x_1' + x_1 x_0) + (x_0 x_2 + x_2 x_0)$$

$$+ (x_0 x_3 + x_0 x_3') + (x_1 x_2 + x_2 x_1') + (x_2 x_3 + x_3' x_2)$$

$$= x_0 + x_2 + x_1 x_3 + x_1' x_3' + x_0 \underbrace{(x_1' + x_1)}_{=1} + \underbrace{(x_0 x_2 + x_2 x_0)}_{=x_0 x_2}$$

$$+ x_0 \underbrace{(x_3 + x_3')}_{=1} + x_2 \underbrace{(x_1 + x_1')}_{=1} + x_2 \underbrace{(x_3 + x_3')}_{=1}$$

$$= x_0 + x_2 + x_1 x_3 + x_1' x_3' + x_0 + x_0 x_2 + x_0 + x_2 + x_2$$

$$= x_0 + x_2 + x_1 x_3 + x_1' x_3' + x_0 + x_0 x_2 \text{ (using the relation } x + x = x)$$

$$= x_0 + x_2 + x_1 x_3 + x_1' x_3' + x_0 \underbrace{(1 + x_2)}_{=1}$$

$$= x_0 + x_2 + x_1 x_3 + x_1' x_3'.$$

The following table gives Boolean expression for each of the other seven segments "b" to "g". The reader is encouraged to imitate the above calculation for segment "a" to verify these expressions.

Segment	Boolean expression
a	$x_0 + x_2 + x_1x_3 + x_1'x_3'$
b	$x_1' + x_2x_3 + x_2'x_3'$
c	$x_1 + x_2' + x_3$
d	$x_0 + x_1'x_2 + x_1x_2'x_3 + x_1'x_3'$
e	$x_1'x_3 + x_2x_3 + x_2x_3'$
f	$x_0 + x_2'x_3' + x_1x_2' + x_1x_3'$
g	$x_0 + x_1x_2' + x_1'x_2 + x_2x_3'$

Using these algebraic expressions for the seven segments, we can now design a logic circuit for the BCD to seven-segment converter.

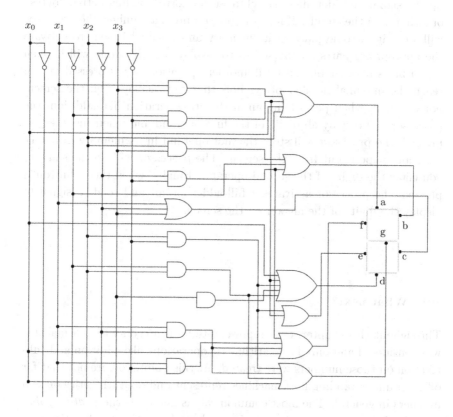

Fig. 1.16   A logic circuit for a BCD to a seven-segment converter.

1.8 So, how does the magic happen?

The logic gates that we encountered in this chapter are in reality mathematical abstractions of electrical devices that vary depending on what type of technology is used. It is easy to find and buy these devices in the market to physically implement a designed logic circuit. Our goal in this chapter was not to go into the technical details of a calculator operation. We hope that with a better understanding of basic binary arithmetics, you would now be interested in learning about the sequence of operations that take place when a specific operation is performed in a calculator. When you press a button, the rubber underneath makes contact with a digital circuit producing an electrical signal in the circuit. The processor of your device picks up the signal and identifies the addresses of corresponding active "bytes" or switches in the circuit. If for example you press a number, the processor will store it in some place in its memory and a signal is sent to activate the appropriate parts on the screen to display it. The same will happen if you press another button until another operation key is pressed or you reach the maximal number of symbols that can displayed on the screen. For example, when performing an arithmetic operation like addition, the processor will display all digits of the first operand and when the + key is pressed, the processor will store the first operand in its memory in binary form and wipe it out from the screen. The processor will do the same as you enter the digits of the second operand. Finally, when the = button is pressed, the processor activates a full-adder circuit and sends a signal to display the digits of the answer on the screen.

1.9 What next?

Throughout this chapter, only representation of integers inside a machine was considered and only basic arithmetic operations like addition and subtraction on those numbers were treated. At this point, you are not that far off from understanding how machines represent and do arithmetics on real numbers in general. The most common way is known as the *floating-point*. You also have now a solid base to learn about how other arithmetic operations (like multiplication, division, exponentiation, comparison and many others) are performed inside a machine.

1.10 References

Brown, S. and Vranesic., Z. (2005). *Digital Logic with VHDL Design, Second Edition.* (McGraw-Hill).

Predko, M. (2005) *Digital electronic demystified.* (McGraw-Hill).

Farhat, H.A. (2004) *Digital design and computer organization.* (CRC Press).

Chapter 2

Basics of data compression, prefix-free codes and Huffman codes

2.1 Introduction

In our era of digital intelligence, data compression became such a necessity that without it our modern lifestyle as we know it will come to a halt. We all use data compression on a daily basis, without realizing it in most cases. Saving a file on your computer, downloading or uploading a picture from or to the internet, taking a photo with your digital camera, sending or receiving a fax or undergoing an MRI medical scan, are few examples of daily activities that require data compression. Without this technology, it would virtually be impossible to do things as simple as viewing a friend's photo album on a social network, let alone complex electronic transactions for businesses and industries. In this chapter, we go over some of the compression recipes where mathematics is the main ingredient.

2.1.1 *What is data compression and do we really need it?*

In the context of this chapter, the word *data* stands for a digital form of information that can be analyzed by a computer. Before it is converted to a digital form, information usually comes in a *raw form* that we call *source data*. Source data can be a text, an image, an audio or a video.

The answer to "why do we need data compression?" is simple: *reduce the cost of using available technologies.* Imagine you own a business for selling or renting moving boxes. If you do not use foldable boxes or collapsible containers, you would need a huge storage facility and your business would not be financially sustainable. Similarly, uncompressed texts, images, audio and video files or information transfer over digital networks require substantial storage capacity certainly not available on standard machines you use at the office or at home.

By a *data compression algorithm*, we usually mean a process through which we can represent data in a compact and digital form that uses less space to store or less time to transmit over a network than the original form. This is usually done by reducing unwanted noise (or redundancy) in the data to a certain degree where it is still possible to recover it in an *acceptable* form. The process of assigning digital codes to pieces of data for storage or transmission is called *encoding*. Of course, a compression algorithm is only efficient when we are able to reverse the process, that is to retrieve the original sequence of characters from the encoded digital form. This process is called *decoding* or *decompressing*. In the literature, the word *compression* often means both the compression and the decompression processes (or encoding and decoding) of data.

2.1.2   *Before you go further*

Mathematical skills like manipulating algebraic inequalities, basic properties of logarithmic functions and some level of discrete mathematics are needed in this chapter.

2.2   Storage inside computers

When you click the "Save" button after working on or viewing a document (text, image, audio,...), a convenient interpretation of what happens next is to imagine that your computer stores the file in the form of a (long) finite sequence of 0's and 1's that we call a binary string. The reason why only two characters are used was explained briefly in the chapter on electronic calculators.

| 0 | 1 | 1 | 1 | 1 | 1 | 1 | 1 | 1 | 1 | 1 | 0 | 0 | 0 | 0 | 1 | 0 | 0 | 0 | 0 | 1 | 0 | 0 | 0 | 0 | 0 | 0 | 0 | 0 | 1 | 0 | 0 | 0 | 1 | 0 | 0 |

In this digital format, each of the two characters "0" and "1" is called a **"bit"** (binary digit). The number of bits in a string is called the *length* of the string. Note that the number of strings of a given length n is 2^n since each bit can take two possible values. For example, there are exactly $2^5 = 32$ different binary strings of length 5.

As the smallest piece of data that can be stored in an electronic device, a bit is too small to deal with or to use in practice. It is much like the penny (1 ¢) which in real life does not buy you anything. Most computers store files in the form of masses of one-dimensional arrays of **bytes**, each byte being a block of 8 bits:

Why 8 bits you may ask? Well, why 12 units in a dozen? But here is a somehow more convincing answer. The famous (extended) ASCII code list (involving symbols on your keyboard plus other characters) contains exactly 256 characters and since $2^8 = 256$, a digital representation for each of the 256 characters in the ASCII code is possible using the byte as the storage unit of one character.

2.2.1 *Measuring units*

Digital files are usually measured by thousands, millions and even billions of bytes. In the standard metric system, the prefix "K" stands for "kilo" or one thousand ($10^3 = 1000$), the prefix "M" stands for "Mega" or one million ($10^6 = 1,000,000$), the prefix "G" stands for "Giga" or one billion ($10^9 = 1,000,000,000$) and so on. The same prefixes are used for sizes of digital files but with different interpretations. Note first that the closest power of 2 to one thousand is 10 since $2^{10} = 1024$, and the closest power of 2 to one million is 20 since $2^{20} = 1,048,576$. In digital sizes, the prefix "KB" stands for "KiloByte" and is equal to $2^{10} = 1024$ bytes (not just 1000 bytes). Similarly, "MB" stands for "MegaByte" and is equal to $2^{20} = 1024 \times 1024 = 1,048,576$ bytes or 1024 KB, "GB" stands for "GigaByte" and is equal to $2^{30} = 1024 \times 1024 \times 1024 = 1,073,741,824$ bytes or 1024 MB, and so on. For example, if a certain digital file has a size of 120KB, a computer uses $120 \times 1024 = 122,880$ bytes to store it.

2.3 Lossy and lossless data compression

Depending on the end goal, data compression algorithms fall into two main categories: *lossless* and *lossy* compressions. Lossless data compression allows the exact original data to be reconstructed from the compressed data. Lossy data compression allows only an "approximation" of the original data to be reconstructed, in exchange for smaller size. Lossless data compression is used in applications where loss of information cannot be tolerated. For example, when compressing a computer program one must make sure that the exact version can be retrieved since the loss of a single character in the program could make it meaningless. Lossy data compression is mainly used for applications where loss of information can be tolerated. For example, in compressing audio, video and still images files, some loss of information could degrade a bit the quality of the decompressed file but not to a degree where human senses can detect a significant difference (see the chapter on JPEG standard).

If lossless compression algorithms exist (and they do), why even bother with lossy compression? Could we not just create an algorithm capable of reducing the size of *any* file and at the same time capable of reconstructing the compressed file to its exact original form? Mathematics gives a definitive answer: Don't bother, such an algorithm is just wishful thinking and cannot exist. First note that a lossless compression algorithm $\mathfrak{C}$ is successful only if the following two conditions are met:

(1) $\mathfrak{C}$ must compress a file of size n bits to a file of size at most $n-1$ bits. Otherwise, no compression has occurred.
(2) If F_1 and F_2 are two distinct files, then their compressed forms $\mathfrak{C}F_1$ and $\mathfrak{C}F_2$ must be distinct as well. Otherwise, we would not be able to reconstruct the original files.

These conditions make it virtually impossible to come up with a compression algorithm that reduces the size of every possible input data file. The following theorem explains why.

Theorem 2.1. There exists no universal algorithm that can compress, in a lossless fashion, all files of a given size n.

Proof. If $k \geq 0$ is an integer, let S_k be the set of all binary strings of length k (files of size k bits). If a universal algorithm α exists, then for each $F \in S_n$ the size of a compressed file $\alpha(F)$ must be at most $n-1$. In other words, if $F \in S_n$, then $\alpha(F) \in S_0 \cup S_1 \cup \cdots \cup S_{n-1}$. But notice that there are exactly 2^i strings in S_i, and the sets are pairwise disjoint (the intersection $S_i \cap S_j$ is empty for $i \neq j$) which means that there are $1 + 2^1 + 2^2 + \cdots + 2^{n-1}$ files in total in the union $S_0 \cup S_1 \cup \cdots \cup S_{n-1}$. The sum $1 + 2^1 + 2^2 + \cdots + 2^{n-1}$ is geometric with n terms and ratio 2.

Therefore,

$$1 + 2^1 + 2^2 + \cdots + 2^{n-1} = \frac{1 - 2^n}{1 - 2} = 2^n - 1.$$

This means that there are $2^n - 1$ different files of size at most $n - 1$. Thus, α must compress at least two elements of S_n (two files of size n) into the same compressed form (of size $n - 1$ at most). This is a contradiction to the second condition of an efficient lossless compression algorithm. □

Another way to interpret the above theorem is the following: for any lossless data compression algorithm $\mathfrak{C}$, there must exist a file that does not get smaller when processed by $\mathfrak{C}$.

2.4   Binary codes

Words in any language are concatenations of basic symbols that we refer to as the alphabet of the language. From a compression perspective, an input file is usually formed by a finite set of symbols $\mathcal{A} = \{\alpha_1, \ldots, \alpha_r\}$ that we call the *alphabet*. By assigning binary representation c_i (a string of 0's and 1's) to each character α_i in the alphabet, the set $\mathcal{C} = \{c_1, c_2, \ldots, c_n\}$ we obtain is called a *binary code* and each c_i is called a *codeword*. For example, if $\mathcal{A} = \{X, Y, Z, W\}$, then $\mathcal{C} = \{00110, 0, 1010, 111\}$ is a binary code for $\mathcal{A}$ with codewords 00110, 0, 1010 and 111.

2.4.1   *Binary trees*

A useful way to represent a binary code is to associate to it a picture called a *binary tree* in the following way. Start the tree with a node (○) at the root. Pick a codeword and read its bits one at a time from left to right. If a bit 0 is read, create a left branch with a new node at its end. Draw a right branch otherwise. Move to the next bit in the codeword. If it is 0, move down one edge along the left branch and record a new node at the end of the new edge. Do the same but along the right branch (from the previous node) if the second bit is 1. Repeat the process until all bits are read. Mark the character after finishing reading all the bits in the codeword. Starting from the root node, repeat the same process for the second codeword. The tree is complete when all codewords are considered. Note that each node can have at most two branches. Each branch represents either 0 or 1. If a node does not grow any more branches, it is called a *leaf*. Otherwise, we call it an *internal node*.

Example 2.1. Consider the binary code $C = \{00, 01, 001, 0010, 1101\}$ for the alphabet $\Sigma = \{X, Y, Z, U, V\}$. The binary tree for C is given below.

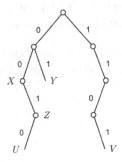

2.4.2 *Fixed length and variable length codes*

The extended version of the ASCII code transforms letters (English alphabet), punctuation, numbers and other symbols (for a total of 256) into binary codes of length 8 each (1 byte per character). For example, the (capital) letter "J" is assigned the codeword 01001010, the letter "o" is assigned the codeword 01101111 and 01100101 is the codeword for the letter "e". The ASCII code is an example of a *fixed length* code since it assigns the same length (8 bits) for each codeword. However, fixed length codes are not usually the most efficient for compression purposes. To save storage space, it would make more sense to assign shorter codewords for characters that appear frequently in a source data and longer ones for less frequent characters. For example, some studies suggest that in a typical English text, the letter "e" appears more frequently than other letters (this is not exactly true if we include other non-letter characters, like the space character which is a bit more frequent than "e") followed by the letter "t", whereas the letters "q" and "z" appear at the bottom of the list of most frequent letters. So, if we are compressing an English text, a good compression strategy would be to assign shorter codewords for "e" and "t" and longer ones for "q" and "z". In this case, our code would be called a *variable length code*. The famous Morse code is an example of a variable length code where just a dot is assigned to the letter "e" and just a dash to the letter "t" (as opposed to combinations of dots and dashes for other letters and symbols).

2.5 Prefix-free code

Variable length codes are very practical for compression purposes, but not enough to achieve optimal results. For a code to be efficient, one must include in the design a unique way of decoding. For example, consider the code $C = \{0, 01, 10, 1001, 1\}$ for the alphabet $\{a, b, c, d, e\}$. The word 101001 can be interpreted in more than one way: "cd", "ead" or "ccb". This is certainly not a well-designed code because of this decoding ambiguity. A closer look at C shows that the problem with it is the fact that some codewords are prefixes (or start) of other codewords. For example, the codeword 10 is a prefix of the codeword 1001. One way to avoid the confusion is to include a symbol to indicate the end of a codeword, but this risks to be costly considering the number of times the end symbol must be included in the encoded file. A code C with the property that no codeword in C is a prefix of any other codeword in C is called a *prefix-free* code. Prefix-free codes are *uniquely decodable* codes in the sense that there is unique way to decode any encoded message. The converse is not true in general. There are examples of uniquely decodable codes which are not prefix-free.

Example 2.2. The code $H = \{10, 01, 000, 1111\}$ is a prefix-free code since no codeword is a start of another codeword. In particular, it is uniquely decodable.

Example 2.3. The code $H = \{00, 011, 10\}$ for the alphabet $\{a, b, c\}$ is uniquely decodable (try to decode couple of binary sequences and you will quickly see why) but clearly not prefix-free.

2.5.1 *Decoding a message using a prefix-free code*

Using a prefix-free code C, decoding a binary message can be achieved using the binary tree of C as follows.

(1) Starting at the root of the tree, move down and left one branch if 0 is the first bit in the encoded message, and move down and right one branch if the first bit is 1.
(2) Repeat using adjacent bits in the sequence until you reach an external leaf node.

(3) Record the letter corresponding to the leaf node.
(4) Return to the root of the tree and repeat the previous steps using other blocks of digits in the encoded sequence.

Example 2.4. Consider the code $C = \{00, 01, 10, 110, 111\}$ for the alphabet $\{A, B, C, D, E\}$. It is clear that C is prefix-free since no codeword is a prefix of another. Suppose we want to decode the binary message 01000011010111. A good way to start is to draw the binary tree associated with C:

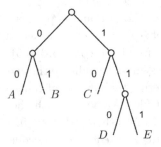

Starting at the root of the tree, we move left using the first 0 in the string and then we move right using the first 1. A leaf with label "B" is reached. We record the letter "B" and return to the root. At this point, we continue with the string 000011010111. From the root, we move left (for the first 0 in the new string) and then down along the same branch (for the second 0 in the new string). The character "A" is reached and we record it next to the letter "B" previously found. Continuing in this manner, we see that the string 01000011010111 corresponds to the message "BAADCE".

2.5.2 *How to decide if a code is prefix-free?*

Deciding if a given code is prefix-free just by looking at the codewords can be challenging. The associated binary tree is one way to simplify this task. Looking at the tree, the code is prefix-free if all the alphabet characters are associated with leaf nodes. If one character happens to be at an internal node, the code is not prefix-free. The code in Example 2.1 is not prefix-free since the characters X and Z are associated with internal leaves. The code in Example 2.4 is prefix-free since every character corresponds to a leaf node in the tree.

2.5.3   The Kraft inequality for prefix-free binary codes

Given a list of positive integers $\mathcal{L} = (l_1, l_2, \ldots, l_n)$, is it possible to construct a prefix-free binary code with $\mathcal{L}$ as codeword lengths? The answer is yes if a certain condition (known as the Kraft inequality) on the lengths l_i is satisfied. Namely, we have the following.

Theorem 2.2 (Kraft inequality). Let $\mathcal{A} = \{\alpha_1, \alpha_2, \ldots, \alpha_n\}$ be an alphabet, $\mathcal{L} = (l_1, l_2, \ldots, l_n)$ a list of n positive integers. A prefix-free binary code $\mathcal{C}$ on $\mathcal{A}$ with $\mathcal{L}$ as codeword lengths exists if and only if

$$\sum_{j=1}^{n} 2^{-l_j} \leq 1. \tag{2.1}$$

Proof. There are several proofs of this well-known result in the literature. Some of them are purely algebraic, others are schematic using trees. Theorem 2.2 consists of a necessary and a sufficient condition. In what follows, we use a schematic proof for one direction and an algebraic one for the other to give the reader a flavor of both approaches.

Assume $\mathcal{C} = \{c_1, c_2, \ldots, c_n\}$ is a prefix-free binary code of the alphabet $\mathcal{A}$. Let l_i be the length of c_i, the codeword corresponding to α_i and let $l = \max\{l_i;\ i = 1, \ldots, n\}$ (l is the largest codeword length). Construct the *complete and full maximal* binary tree T of height l. That is to say, T is the binary tree in which every internal node has exactly two children nodes (two branches) with every level having the maximum possible number of nodes, and leaves in the tree appear only at the deepest level l. For example, the following is the complete and full maximal binary tree T of height 4:

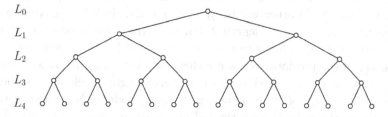

Note that T has exactly 2^l leaves, which correspond to all possible codewords of length l. The length l_i is called the *depth* of α_i in the tree and c_i is formed by collecting all the binary symbols on the path from the root to α_i. Next, we extract the binary tree $T(\mathcal{C})$ corresponding to $\mathcal{C}$ as a "subtree" of T. Since the code $\mathcal{C}$ is prefix-free, we know that every character

α_i corresponds to a leaf in $T(\mathcal{C})$. The fact that character α_i is at depth l_i implies that there is a total of 2^{l-l_i} codewords with c_i as prefix. These 2^{l-l_i} codewords must be erased in order for α_i to correspond to a leaf. Doing this for every $i = 1, 2, \ldots, n$, the total number of ruled out codewords is $\sum_{i=1}^{n} 2^{l-l_i}$. But this number cannot possibly exceed 2^l. Therefore

$$\sum_{i=1}^{n} 2^{l-l_i} \leq 2^l \implies \sum_{i=1}^{n} 2^l 2^{-l_i} \leq 2^l \implies \sum_{j=1}^{n} 2^{-l_j} \leq 1.$$

This proves one direction in the Theorem.

For the other direction, assume that the inequality $\sum_{j=1}^{n} 2^{-l_j} \leq 1$ is satisfied for a certain set of positive integers $\mathcal{L} = \{l_1, \ldots, l_n\}$. We need to construct a prefix-free code $\mathcal{C}$ of the alphabet $\mathcal{A}$ with the l_i's as codeword lengths. For each positive integer k, let $\mathcal{L}_k$ be the subset of $\mathcal{L}$ consisting of those l_j's equal to k and let γ_k be the number of elements in $\mathcal{L}_k$. For example, if $\mathcal{L} = \{1, 2, 3, 5, 5, 5, 5\}$, then $\mathcal{L}_1 = \{l_1\}$, $\mathcal{L}_2 = \{l_2\}$, $\mathcal{L}_3 = \{l_3\}$, $\mathcal{L}_4 = \emptyset$ (empty), $\mathcal{L}_5 = \{l_4, l_5, l_6, l_7\}$ and $\gamma_1 = 1$, $\gamma_2 = 1$, $\gamma_3 = 1$, $\gamma_4 = 0$ and $\gamma_5 = 4$. For any $k \notin \{1, 2, 3, 4, 5\}$, $\mathcal{L}_k = \emptyset$ and $\gamma_k = 0$. Note that by regrouping like powers of 2 in the left side of inequality (2.1), the inequality can be written as

$$\sum_{k=1}^{l} \gamma_k 2^{-k} \leq 1, \tag{2.2}$$

where $l = \max\{l_i; i = 1, \ldots, n\}$. Starting with $\mathcal{L}_1$, the subset of code length 1, it is clear that we can choose at most two codewords of length 1, namely 0 and 1. In other words, $\gamma_1 \leq 2$. For $\mathcal{L}_2$, there is a maximum of 2^2 (distinct) codewords of length 2, but since the desired code $\mathcal{C}$ must be prefix-free, we must eliminate those codewords of length 2 having one or the other of the previously chosen codewords. This could be done as long as $\gamma_2 \leq 2^2 - 2\gamma_1$. To create the γ_3 codewords of length 3 each, none of these codewords (a total of 2^3) can start with the γ_1 codewords created at the first step or the γ_2 codewords created at the second step. There are $2^2\gamma_1$ codewords of length 3 having prefixes from the codewords created at the first step and $2\gamma_2$ codewords of length 3 having prefixes from the codewords created at the second step. Therefore γ_3 must satisfy the inequality $\gamma_3 \leq 2^3 - 2^2\gamma_1 - 2\gamma_2$. Continuing this process, we get that the desired prefix-free code is possible provided that the following system of inequalities is satisfied:

$$(S) \begin{cases} \gamma_1 \leq 2 \\ \gamma_2 \leq 2^2 - 2\gamma_1 \\ \gamma_3 \leq 2^3 - 2^2\gamma_1 - 2\gamma_2 \\ \gamma_4 \leq 2^4 - 2^3\gamma_1 - 2^2\gamma_2 - 2\gamma_3 \\ \vdots \quad \vdots \quad \vdots \\ \gamma_l \leq 2^l - 2^{l-1}\gamma_1 - 2^{l-2}\gamma_2 - \cdots - 2\gamma_{l-1} \end{cases}.$$

Note that if the second inequality in (S) is satisfied, then $2\gamma_1 \leq 2^2 - \gamma_2$ and $\gamma_1 \leq 2 - \frac{\gamma_2}{2}$. But since $2 - \frac{\gamma_2}{2} \leq 2$, we get that $\gamma_1 \leq 2$ and the first inequality in (S) is also satisfied. Similarly, it is easy to see that if any inequality in (S) is satisfied, then the previous one is also satisfied. Therefore, if we can prove that the last inequality is satisfied, then the system (S) would be valid and a prefix-free binary code can be constructed. Multiplying the last inequality in (S) by 2^{-l} and rearranging gives:

$$2^{-1}\gamma_1 + 2^{-2}\gamma_2 + \cdots + 2^{-(l-1)}\gamma_{l-1} \leq 1.$$

This is the same as the inequality (2.2) above. We conclude that a prefix-free code can be constructed if the Kraft inequality is satisfied. $\qquad\square$

Remark 2.1. It is important to have a clear understanding of what Kraft's inequality says about the code and more importantly, of what it does not say. The inequality serves as a tool to (quickly) check if a given binary code is *not* prefix-free. However, it does offer much help to prove that a given code is indeed prefix-free. For example, the binary code $\mathcal{C}_1 = \{10, 11, 01, 00, 0111\}$ is not a prefix-free since the codewords lengths do not satisfy Kraft's inequality, so there is no point of wasting time to check if there is a codeword which is a prefix of another, we know it is the case. On the other hand, the codeword lengths of $\mathcal{C}_2 = \{10, 101, 01\}$ satisfy Kraft's inequality, but $\mathcal{C}_2$ is clearly not prefix-free.

2.6   Optimal codes

Consider an alphabet with n symbols $\mathcal{A} = \{\alpha_1, \ldots, \alpha_n\}$. Assume that a certain source file with alphabet $\mathcal{A}$ has a *probability distribution vector* $\mathcal{P} = (p_1, \ldots, p_n)$. This means that p_i is the probability of occurrence of the symbol α_i in the file. We can think of the probability p_i as the ratio of the number of times α_i appears in the file by the total number of occurrences of all alphabet symbols. A consequence of this interpretation is that $\sum_{i=1}^{n} p_i = 1$. For a prefix-free binary code of the alphabet $\mathcal{A}$ with codeword lengths $\mathcal{L} = \{l_1, \ldots, l_n\}$, we define the *average codeword length* of $\mathcal{C}$ as being $l_{av} = \sum_{i=1}^{n} p_i l_i$ measured in bits per source symbols.

We make the following observations without looking too much into the probabilistic and statistical properties of the source.

(1) It is completely irrelevant what names we give for the source alphabet symbols. All that matters at the end of the day is the probability distribution vector of these symbols.

(2) If $\mathcal{C}$ is a prefix-free binary code of $\mathcal{A}$ with codeword lengths $\mathcal{L} = \{l_1, \ldots, l_n\}$, then the average codeword length $l_{av} = \sum_{i=1}^{n} p_i l_i$ of $\mathcal{C}$ can be thought of as the **average number of bits per symbol** required to encode the source.

(3) It is then natural to seek binary prefix-free codes with average codeword length as small as possible in order to save on the numbers of bits used to encode the source. A prefix-free code with minimal average codeword length is called an *optimal code*.

Example 2.5. Consider the source file $EBRRACCADDABRA$ with alphabet $\mathcal{A} = \{A, B, C, D, E, R\}$. The source probability distribution vector is $\mathcal{P} = \left(\frac{2}{7}, \frac{1}{7}, \frac{1}{7}, \frac{1}{7}, \frac{1}{14}, \frac{3}{14}\right)$. For the binary (prefix-free) code

$$\mathcal{H} = \{1010, 001, 111, 100, 01, 110\}$$

of the alphabet, the average codeword length is $l_{av} = \left(\frac{2}{7}\right)(4) + \left(\frac{1}{7}\right)(3) + \left(\frac{1}{7}\right)(3) + \left(\frac{1}{7}\right)(3) + \left(\frac{1}{14}\right)(2) + \left(\frac{3}{14}\right)(3) = \frac{45}{14} \cong 3.2$. This means that it takes, on average, 3.2 bits per alphabet symbol to encode the source using $\mathcal{H}$. Note that the code $\mathcal{H}$ is not optimal since the (prefix-free) code

$$\mathcal{C} = \{00, 010, 011, 100, 101, 11\}$$

of the same alphabet has an average codeword length of $\left(\frac{2}{7}\right)(2) + \left(\frac{1}{7}\right)(3) + \left(\frac{1}{7}\right)(3) + \left(\frac{1}{7}\right)(3) + \left(\frac{1}{14}\right)(3) + \left(\frac{3}{14}\right)(2) = 2.5$ bits per source symbols. So it takes fewer bits to encode the source using $\mathcal{C}$.

It is not entirely obvious that an optimal code exists for a given alphabet. The following result proves that in fact an optimal and prefix-free binary code always exists.

Theorem 2.3. Given a source $\mathcal{S}$ with alphabet $\mathcal{A} = \{\alpha_1, \ldots, \alpha_n\}$ and a probability distribution vector $\mathcal{P} = (p_1, \ldots, p_n)$, an optimal prefix-free binary code for $\mathcal{S}$ exists.

Proof.   Without any loss of generality, we can assume that $p_i > 0$ for each $i = 1, \ldots, n$. Indeed, if this is not the case, we can restrict our alphabet to those symbols with positive probability and just ignore all symbols with zero probability (they do not appear in the source anyway). By rearranging the symbols if necessary, we can also assume that $p_1 \leq \cdots \leq p_n$. We start by proving the existence of a prefix-free binary code for $\mathcal{A}$. This can be achieved in more than one way. First, take l_i to be the smallest integer greater than or equal to $-\log_2(p_i)$ for each $i = 1, \ldots, n$. The proof of Theorem 2.4 below shows in particular that the list $(l_1, \ldots, l_n)$ satisfies the Kraft inequality and hence represents the codeword lengths of a prefix-free binary code of the alphabet. Another way to construct a prefix-free code on the alphabet $\mathcal{A}$ is to choose a positive integer l satisfying $2^l \geq n$, then the integers $l_1 = l_2 = \cdots = l_n = l$ satisfy:

$$\sum_{i=1}^{n} 2^{-l_i} = n2^{-l} = \frac{n}{2^l} \leq 1.$$

By Theorem 2.2, $(l_1, \ldots, l_n)$ is the codeword lengths of a prefix-free code on $\mathcal{A}$. Next, let us fix a prefix-free code $\mathcal{C}_0$ on $\mathcal{A}$ of average codeword length l_0. We claim that there is only a finite number of binary prefix-free codes with average codeword length less than or equal to l_0. To see this, let $\mathcal{C}$ be a prefix-free code on $\mathcal{A}$ of codeword lengths $(l_1, \ldots, l_n)$ and an average codeword length $l = \sum_{i=1} p_i l_i$ satisfying $l \leq l_0$. If $l_k > \frac{l_0}{p_1}$ for some k, then

$$l = \sum_{i=1}^{n} p_i l_i \geq p_k l_k > p_1 \frac{l_0}{p_1} = l_0$$

which contradicts the assumption $l \leq l_0$. We conclude that $l_k \leq \frac{l_0}{p_1}$ for $k = 1, \ldots, n$. Clearly, there are finitely many codewords of length less than or equal to the constant $\frac{l_0}{p_1}$ and thus the set $\mathcal{G}$ of all binary prefix-free codes with average codeword length less than or equal to l_0 is finite. Pick a code $\mathcal{C}$ in $\mathcal{G}$ with the lowest average codeword length (this is possible since $\mathcal{G}$ is finite), then $\mathcal{C}$ is a prefix-free and optimal code for the alphabet.   □

Given a source with alphabet $\mathcal{A} = \{\alpha_1, \ldots, \alpha_n\}$ and a probability distribution vector $\mathcal{P} = (p_1, \ldots, p_n)$, let $\bar{l}$ be the minimum of the set

$$\left\{ \sum_{i=1}^{n} p_i l_i; \ l_i \text{ is a positive integer and } \sum_{i=1}^{n} 2^{-l_i} \leq 1 \right\}.$$

In other words, $\bar{l}$ is the minimal average codeword length taken over all possible prefix-free codes of the source. Theorem 2.3 guarantees the existence of a prefix-free code with average codeword length equals to $\bar{l}$. But how can we actually construct that code? The Kraft inequality gives us a clean and nice mathematical formulation of the problem at hand.

Problem 2.1. If $\mathcal{P} = (p_1, \ldots, p_n)$ is a given probability distribution vector (i.e., $\sum_{i=1}^{n} p_i = 1$ and $p_i > 0$ for $i = 1, \ldots, n$), how can we find n positive integers $l_1, \ldots, l_n$ subject to the constraint $\sum_{i=1}^{n} 2^{-l_i} \leq 1$ (Kraft inequality) so that the sum $\sum_{i=1}^{n} p_i l_i$ is minimal? In other words, how can we find positive integers $l_1, \ldots, l_n$ satisfying the Kraft inequality so that $\sum_{i=1}^{n} p_i l_i = \bar{l}$?

Problem 2.1 is a classic optimization problem involving several variables $(l_1, \ldots, l_n)$ subject to a constraint (the Kraft inequality). In addition, we have another complication to deal with: the variables l_i's must be (positive) integers. If we ignore that last constraint on the variables and assume that each l_i is just a real number, the problem can be dealt with quite efficiently using a technique known as the *method of Lagrange multipliers*. Readers familiar with this optimization technique are encouraged to try to apply it to this particular problem for small values of n. We will not go over the solution of Problem 2.1 in this book. However, we present an interesting result that gives a lower and an upper bound on the value of $\bar{l}$.

Theorem 2.4. Let $\mathcal{P} = (p_1, \ldots, p_n)$ be the probability distribution vector of a certain data source with alphabet $\mathcal{A} = \{\alpha_1, \ldots, \alpha_n\}$. If $\bar{l}$ is the average codeword length of any *optimal* prefix-free code for the source, then

$$-\sum_{i=1}^{n} p_i \log_2 (p_i) \leq \bar{l} < 1 - \sum_{i=1}^{n} p_i \log_2 (p_i). \tag{2.3}$$

Moreover, there exists a prefix-free binary code with codeword lengths $(l_1, \ldots, l_n)$ and average codeword length equals to $-\sum_{i=1}^{n} p_i \log_2 (p_i)$ if and only if $p_i = 2^{-l_i}$ for each $i = 1, \ldots, n$.

Proof. For the proof of the inequality on the left in (2.3), we use a well-known inequality that you probably have seen in your first Calculus course:

$$\ln(x) \leq x - 1, \quad \text{for } x > 0 \tag{2.4}$$

with equality in (2.4) occurring only at $x = 1$. This can be seen from the graphs of both $\ln(x)$ and $x - 1$:

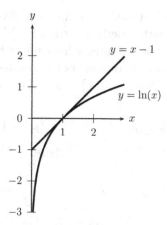

Since $\ln(x) = \frac{\log_2(x)}{\log_2(e)}$ (change of base for logarithmic functions), inequality (2.4) can be written as

$$\log_2(x) \leq \log_2(e)(x-1), \quad \text{for } x > 0. \tag{2.5}$$

Given any prefix-free binary code $\mathcal{C}$ of the alphabet $\mathcal{A}$ with codeword lengths $(l_1, \ldots, l_n)$ and average codeword length l_{av}, we have:

$$-\sum_{i=1}^{n} p_i \log_2(p_i) - l_{av}$$

$$= -\sum_{i=1}^{n} p_i \log_2(p_i) - \sum_{i=1}^{n} p_i l_i = -\sum_{i=1}^{n} p_i \left(\log_2(p_i) + l_i\right)$$

$$= -\sum_{i=1}^{n} p_i \left(\log_2(p_i) + \log_2(2^{l_i})\right) = -\sum_{i=1}^{n} p_i \log_2\left(p_i 2^{l_i}\right)$$

$$= \sum_{i=1}^{n} p_i \log_2\left(p_i^{-1} 2^{-l_i}\right) = \sum_{i=1}^{n} p_i \log_2\left(\frac{2^{-l_i}}{p_i}\right)$$

$$\leq \sum_{i=1}^{n} p_i \log_2(e) \left(\frac{2^{-l_i}}{p_i} - 1\right) \quad \text{(by (2.5) since } \frac{2^{l_i}}{p_i} > 0\text{)} \quad (*)$$

$$= \log_2(e) \sum_{i=1}^{n} \left(2^{-l_i} - p_i\right) = \log_2(e) \left[\sum_{i=1}^{n} 2^{-l_i} - \sum_{i=1}^{n} p_i\right]$$

$$\leq 0 \text{ (since } \sum_{i=1}^{n} 2^{-l_i} \leq 1 \text{ by Kraft and } \sum_{i=1}^{n} p_i = 1\text{)}.$$

We conclude that the average codeword length l_{av} of any prefix-free code satisfies the inequality $-\sum_{i=1}^{n} p_i \log_2(p_i) \leq l_{av}$. Since any optimal code is in particular prefix-free, the first inequality in (2.3) is established. For the second (strict) inequality, it suffices to find a prefix-free binary code with an average codeword length strictly less than $1 - \sum_{i=1}^{n} p_i \log_2 (p_i)$. To this end, let $l_i = \lceil - \log_2(p_i) \rceil$ be the integer least upper bound of $- \log_2(p_i)$ for $i = 1, \ldots, n$. That is to say, l_i represents the smallest integer greater than or equal to $- \log_2(p_i)$. Since $x \leq \lceil x \rceil < x + 1$ for any real number x, we have

$$- \log_2(p_i) \leq l_i < - \log_2(p_i) + 1 \tag{2.6}$$

or equivalently $\log_2(p_i) - 1 < -l_i \leq \log_2(p_i)$. This gives the following inequalities

$$2^{(\log_2(p_i)-1)} < 2^{-l_i} \leq 2^{\log_2(p_i)} \iff \frac{p_i}{2} < 2^{-l_i} \leq p_i.$$

In particular, $\sum_{i=1}^{n} 2^{-l_i} \leq \sum_{i=1}^{n} p_i = 1$. The Kraft inequality is then satisfied. Theorem 2.2 guarantees the existence of a binary prefix-free code with l_i's as codeword lengths. Using (2.6),

$$\sum_{i=1}^{n} p_i l_i < \sum_{i=1}^{n} p_i \left(- \log_2(p_i) + 1 \right)$$

$$= - \sum_{i=1}^{n} p_i \log_2(p_i) + \sum_{i=1}^{n} p_i = - \sum_{i=1}^{n} p_i \log_2(p_i) + 1.$$

This finishes the proof of (2.3) in the Theorem. For the last statement, assume that $\mathcal{C}$ is a prefix-free binary code with codeword lengths $(l_1, \ldots, l_n)$ and average codeword length equals to $- \sum_{i=1}^{n} p_i \log_2 (p_i)$. In order to achieve $l_{av} = - \sum_{i=1}^{n} p_i \log_2 (p_i)$, inequality labeled $(*)$ in the above proof must be an equal sign. That can only happen when $\frac{2^{-l_i}}{p_i} = 1$ for each i (remember that inequality (2.5) is an equal sign if and only if $x = 1$) and so $p_i = 2^{-l_i}$ for all i. Conversely, if $(l_1, \ldots, l_n)$ are positive integers satisfying $p_i = 2^{-l_i}$ for all i, then $-l_i = \log_2 p_i$ and $\sum_{i=1}^{n} 2^{-l_i} = \sum_{i=1}^{n} 2^{\log_2 p_i} = \sum_{i=1}^{n} p_i = 1$. The Kraft inequality implies that there exists a prefix-free binary code $\mathcal{D}$ with codeword lengths $(l_1, \ldots, l_n)$. Note that the average codeword length of $\mathcal{D}$ is $\sum_{i=1}^{n} p_i l_i = - \sum_{i=1}^{n} p_i \log_2 (p_i)$. $\square$

2.7   The source entropy

Given a data source with probability distribution vector $\mathcal{P} = (p_1, \ldots, p_n)$ and alphabet $\mathcal{A}$, the lower bound $H = - \sum_{i=1}^{n} p_i \log_2(p_i)$ on the average codeword length of a prefix-free code of $\mathcal{A}$ shown in Theorem 2.4 seems to fall from the sky. But readers who carried out the details of the Lagrange multipliers technique suggested earlier would have probably seen it popping up in their calculations. It turns out that H plays a central role in information theory, where it is known as the *Entropy* of the source.

In terms of the entropy, two important facts about the source can be drawn from Theorem 2.4. The first is that the average codeword length of any optimal code cannot get any better than the entropy $H = -\sum_{i=1}^{n} p_i \log_2(p_i)$ of the source while it is always within only one bit of the entropy. The second fact is that there exists a prefix-free code of a source with probability distribution vector $\mathcal{P} = (p_1, \ldots, p_n)$ that achieves the entropy bound if and only if the source is dyadic, that is each p_i is of the form $p_i = 2^{-l_i}$ for some positive integer l_i.

Example 2.6. Assume that the probability distribution vector of a source file is $\mathcal{P} = (0.2, 0.1, 0.4, 0.3)$. Assume also that the prefix-free code $\mathcal{C} = \{1, 01, 001, 000\}$ is used to encode the source file. The entropy of the source is

$$H = -\sum_{i=1}^{n} p_i \log_2(p_i)$$
$$= -\left[0.2 \log_2(0.2) + 0.1 \log_2(0.1) + 0.4 \log_2(0.4) + 0.3 \log_2(0.3)\right]$$
$$\approx 1.85 \text{ bits.}$$

This means that, on average, the source code requires a minimum of 1.85 bits to encode. Note that the average codeword length of the code $\mathcal{C}$ is

$$l_{av} = \sum_{i=1}^{n} p_i l_i = (0.2)(1) + (0.1)(2) + (0.4)(3) + (0.3)(3) = 2.5$$

bits per symbol. The fact that the average codeword length of $\mathcal{C}$ is not equal to the entropy of the source is due to the fact that the source is not dyadic.

2.8   The Huffman code

In the late 1940's, researchers in the (then) young field of information theory worked hard on the problem of constructing optimal codes with not much luck. Some descriptions of what such a code should look like were given but without any concrete algorithm to construct one. In the early 1950's, David Huffman was a student in a graduate course at MIT given by Robert Fano on information theory. Huffman and his classmates were given the option to submit a paper on the optimal code question, a problem Fano and others had almost given up hope on solving it, or to write a standard final exam in the course. Huffman worked on the paper for a period of time and just before giving up and going back to study for the final exam, a solution hit him. To everyone's surprise, Huffman's paper consisted of a simple and straightforward way to construct the optimal code and earned him a great deal of fame. While almost every attempted solution to the problem consisted of constructing a code tree from the top down, Huffman's approach was to construct the tree of his code from the bottom up.

Fix a source file with an alphabet $\mathcal{A} = \{\alpha_1, \ldots, \alpha_n\}$ and a probability distribution vector $\mathcal{P} = (p_1, \ldots, p_n)$. Huffman's construction of an optimal prefix-free code is based on the following observations.

Observation 1. *In the binary tree corresponding to an optimal prefix-free code of $\mathcal{A}$, each internal node must have two children* (such a tree is called *full* in the literature). To see this, assume to the contrary that there exists an internal node v with a unique child w. Move w up one level to its parent, that is merge v and w into a unique node. The resulting tree will remain that of a prefix-free code of $\mathcal{A}$ except its average codeword length is shorter than the original one. This is a contradiction to the optimality of the code.

Observation 2. *For an optimal prefix-free code of $\mathcal{A}$ with codeword lengths $l_1, \ldots, l_n$, if $p_i > p_j$ then $l_i \leq l_j$.* This observation should come as no surprise as we expect an optimal code to assign shorter lengths to more probable characters and longer ones to least probable characters. For the proof, assume to the contrary that $\mathcal{C}$ is an optimal prefix-free code satisfying $l_i > l_j$ with $p_i > p_j$ for some $i \neq j$. The average codeword length of $\mathcal{C}$ contains the term $p_i l_i + p_j l_j$. Let $\mathcal{C}'$ be the binary code obtained by interchanging the two codewords corresponding to α_i and α_j. The code $\mathcal{C}'$ is clearly prefix-free and its average codeword length is the same as that of $\mathcal{C}$ with the exception that the term $p_i l_i + p_j l_j$ is replaced by $p_i l_j + p_j l_i$. Since $(p_i l_i + p_j l_j) - (p_i l_j + p_j l_i) = (p_i - p_j)(l_i - l_j) > 0$, the average codeword of $\mathcal{C}'$ is smaller than that of $\mathcal{C}$, contradicting the optimality of $\mathcal{C}$.

Observation 3. *In an optimal code $\mathcal{C}$ with maximum codeword length l, if c is a codeword of length l, then there exists another codeword c' of length l such that c and c' differ only in their last bit.* To see this, write $c = \gamma_1 \gamma_2 \ldots \gamma_{l-1} \gamma_l$, where each γ_i is a bit. By Observation 1 above, the internal node corresponding to $\gamma_1 \gamma_2 \ldots \gamma_{l-1}$ must have two children. One of the children is the leaf corresponding to the codeword c and the other corresponds to $c' = \gamma_1 \gamma_2 \ldots \gamma_{l-1} \bar{\gamma}_l$ where $\bar{\gamma}_l$ is 0 if γ_l is 1 and $\bar{\gamma}_l$ is 1 if γ_l is 0. Since c' is of maximal length l, it must correspond to a leaf in the tree. This implies that c' is also a codeword of $\mathcal{C}$ of maximal length l and it differs from c only in the last bit.

Putting together the above observations leads to the following result.

Theorem 2.5. Consider a source file with alphabet $\mathcal{A} = \{\alpha_1, \ldots, \alpha_n\}$ and probability distribution vector $\mathcal{P} = (p_1, \ldots, p_n)$ with the probabilities arranged in non-decreasing order $p_1 \leq p_2 \leq \cdots \leq p_n$. Then, there exists an optimal prefix-free code of $\mathcal{A}$ satisfying the following property: the codewords corresponding to α_1 and α_2 (the two least probable symbols) have the same **maximal** length and they differ only in their last bit.

Proof.   Start with an arbitrary *optimal* code $\mathcal{C} = \{c_1, \ldots, c_n\}$ of $\mathcal{A}$ of codeword lengths $(l_1, \ldots, l_n)$ where c_i is the codeword assigned to symbol α_i and l_i is the length of c_i (such a code exists by Theorem 2.3). By Observation 2, codeword c_1 has a maximal length l_1 since α_1 has the smallest occurrence probability. By Observation 3, there exists a codeword c_k of the same (maximal) length l_1 and which differs from c_1 only in the last bit. If $k = 2$, we are done. If $k \geq 3$, then $p_k \geq p_3 \geq p_2$ by the ordering on the probabilities chosen above. On the other hand, since $l_k = l_1 \geq l_2$, Observation 2 implies that $p_k \leq p_2$. We conclude that $p_k = p_2$. Let $\mathcal{C}'$ be the code obtained from $\mathcal{C}$ by interchanging codewords α_2 and α_k, then $\mathcal{C}'$ remains prefix-free and optimal since it has the same average codeword length as $\mathcal{C}$. Clearly, $\mathcal{C}'$ satisfies the required property of the theorem.   □

2.8.1   *The construction*

Consider a source with alphabet $\mathcal{A} = \{\alpha_1, \ldots, \alpha_n\}$ and probability distribution vector $\mathcal{P} = (p_1, \ldots, p_n)$. Assume that $\mathcal{C} = \{c_1, \ldots, c_n\}$ is the optimal prefix-free code for $\mathcal{A}$ with codeword lengths $l_1, \ldots, l_n$ and average codeword length l that satisfies the property of Theorem 2.5. Let $\mathcal{T}$ be the binary tree associated with $\mathcal{C}$. If α_i and α_j are the two least probable characters in $\mathcal{A}$, Theorem 2.5 tells us that c_i, c_j have the same (maximal) length and that they differ only in their last bits. For the tree $\mathcal{T}$, this means that α_i and α_j correspond to sibling leaves in the tree (they have the same parent node). Merge the two siblings into a common leaf a_{ij} placed at their parent node in $\mathcal{T}$. Let c_{ij} be the codeword corresponding to a_{ij}; that is, c_{ij} is the codeword consisting of the first common $l_i - 1$ bits of c_i and c_j. The new tree we obtain corresponds to a prefix-free code $\mathcal{C}'$ for the alphabet $\mathcal{A}'$ obtained from $\mathcal{A}$ by removing the symbols α_i, α_j and replacing them with the common symbol α_{ij} to which we assign the probability $p_i + p_j$. If l' is the average codeword length of the code $\mathcal{C}'$, then

$$
\begin{aligned}
l - l' &= p_1 l_1 + \cdots + p_i l_i + \cdots + p_j l_j + \cdots + p_n l_n \\
&\quad - [p_1 l_1 + \cdots + (p_i + p_j)(l_i - 1) + \cdots + p_n l_n] \\
&= p_i l_i + p_j \underbrace{l_j}_{=l_i} - (p_i + p_j)(l_i - 1) = p_i + p_j.
\end{aligned}
$$

In particular, $l = l' + (p_i + p_j)$. The difference between the two average lengths depends solely on the source probability vector. This proves the following lemma.

Lemma 2.1. Let $\mathcal{A} = \{\alpha_1, \ldots, \alpha_n\}$ be an alphabet with a probability distribution vector $\mathcal{P} = (p_1, \ldots, p_n)$. Let $\mathcal{A}'$ be the alphabet obtained from $\mathcal{A}$ by replacing the two least frequent characters α_i and α_j with a single character α_{ij} with assigned probability $p_i + p_j$. If $\mathcal{T}'$ is a binary tree representing an optimal prefix-free code for $\mathcal{A}'$, then the tree $\mathcal{T}$ obtained from $\mathcal{T}'$ by replacing α_{ij} with an internal node with two children α_i, α_j corresponds to an optimal prefix-free code for the original alphabet $\mathcal{A}$.

2.8.2 *The Huffman algorithm*

With all the above results in mind, we are now ready to give a practical algorithm to construct an optimal code due to Huffman. The setup is as before, namely an alphabet $\mathcal{A} = \{\alpha_1, \ldots, \alpha_n\}$ and probability distribution vector $\mathcal{P} = (p_1, \ldots, p_n)$.

(1) Pick two letters α_i and α_j from the alphabet with the smallest probabilities.
(2) Create a subtree with root labeled α_{ij} that has α_i and α_j as leaves.
(3) Set the probability of α_{ij} as $p_i + p_j$.
(4) Form a new alphabet $\mathcal{A}'$ of $n - 1$ symbols by removing α_i and α_j from the alphabet $\mathcal{A}$ and adding the new symbol α_{ij}.
(5) Repeat the previous steps for the new alphabet $\mathcal{A}'$.
(6) Stop when an alphabet with only one symbol is left.

The tree we obtain at the end of the above algorithm is called the Huffman tree and the corresponding code is called the *Huffman code*.

Theorem 2.6. The Huffman code is an optimal prefix-free code.

Proof. We use a proof by induction on n, the number of symbols of the alphabet. If $n = 2$, the Huffman algorithm assigns 0 to one symbol of the alphabet and 1 to the other. Clearly, this is an optimal prefix-free code in this case. Let $n \geq 3$ and assume that the Huffman code returns an optimal prefix-free code for any alphabet of size $n - 1$. Let $\mathcal{A} = \{\alpha_1, \alpha_2, \ldots, \alpha_n\}$ be an alphabet of size n with symbol probabilities arranged in a non-decreasing order $p_1 \leq p_2 \leq \cdots \leq p_{n-1} \leq p_n$. We need to show that Huffman coding returns an optimal prefix-free code for $\mathcal{A}$. Let $\mathcal{A}'$ be the alphabet with $n - 1$ symbols obtained from $\mathcal{A}$ by replacing the two least frequent symbols α_1, α_2 with a single symbol α_{12} with assigned probability $p_1 + p_2$. By the induction hypothesis, the Huffman code will produce an optimal code for $\mathcal{A}'$. Lemma 2.1 shows now that the Huffman code is optimal for $\mathcal{A}$. $\square$

2.8.3 *An example*

In this section, we look at a detailed example of text compression using Huffman algorithm. Assume that we want to encode the following text source:

i see eye in sky

The alphabet of the text is $\mathcal{A} = \{i, s, e, y, n, k, \_\}$, where the symbol $\_$ represents the "space" between words. The symbol probabilities in the text are as follows: $i\left(\frac{1}{8}\right)$, $s\left(\frac{1}{8}\right)$, $e\left(\frac{1}{4}\right)$, $y\left(\frac{1}{8}\right)$, $n\left(\frac{1}{16}\right)$, $k\left(\frac{1}{16}\right)$, and $\_\left(\frac{1}{4}\right)$. We start by arranging the characters according to a non-decreasing order of their probabilities.

$$n(1/16) \quad k(1/16) \quad i(1/8) \quad s(1/8) \quad y(1/8) \quad e(1/4) \quad \_(1/4)$$

The two letters of lowest probabilities in the text are n and k. Create a subtree with root labeled nk to which we assign the with probability $\frac{1}{16} + \frac{1}{16} = \frac{1}{8}$ with n and k as leaves. As usual, the left branch is labeled with a 0 and the right branch is labeled with a 1.

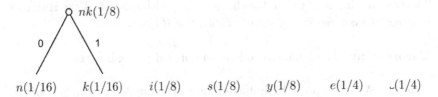

We have now a new alphabet with symbols nk, i, s, y, e and $\_$ and probabilities $\frac{1}{8}$, $\frac{1}{8}$, $\frac{1}{8}$, $\frac{1}{8}$, $\frac{1}{4}$ and $\frac{1}{4}$, respectively. Pick two symbols in the new alphabet of lowest probabilities, say nk and i, and create a new subtree with root labeled nki having nk and i as children and with probability $\frac{1}{8} + \frac{1}{8} = \frac{1}{4}$.

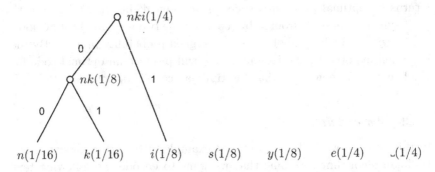

We get the new alphabet formed of symbols nki, s, y, e and $\_$ and probabilities $\frac{1}{4}$, $\frac{1}{8}$, $\frac{1}{8}$, $\frac{1}{4}$ and $\frac{1}{4}$, respectively. Form a new subtree with root labeled sy (probability $\frac{1}{8} + \frac{1}{8} = \frac{1}{4}$) having s and y as children.

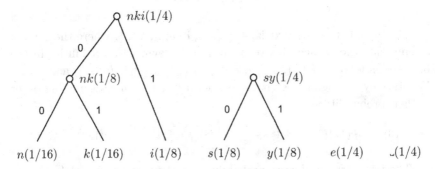

We are now left with an alphabet with four symbols, each occurring with a probability $\frac{1}{4}$. Pick the symbols nki and sy to form the next subtree with node $nkisy$ of probability $\frac{1}{2}$.

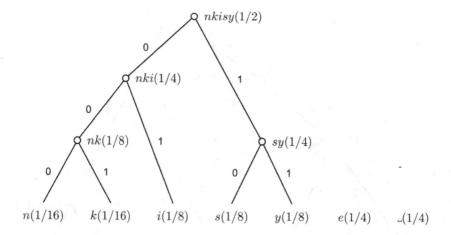

Pick the symbols *e* and the space symbol to form the next subtree.

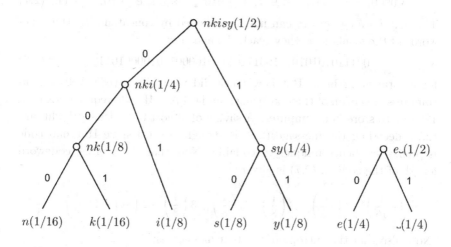

Finally, we merge the last two symbols *nkisy* and *e⌣* into the root of the Huffman tree.

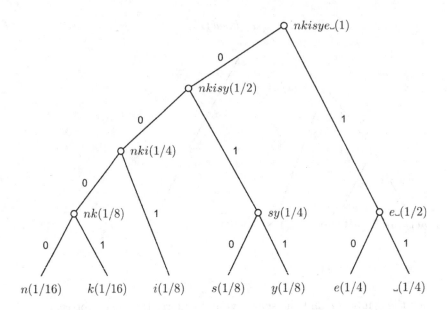

Following the tree from the root to the leaves, we get the *Huffman code* of the text message:

$$n \rightarrow 0000, \ k \rightarrow 0001, \ i \rightarrow 001, \ s \rightarrow 010, \ y \rightarrow 011, \ e \rightarrow 10, \ \text{\textvisiblespace} \rightarrow 11. \quad (2.7)$$

The text *i see eye in sky* can now be encoded by concatenating the codewords of the symbols as they reach the encoder:

$$001110101010111001110110010000110100001011, \quad (2.8)$$

for a total of 42 bits. But how much did we really save? Well, a non-compressed version of the text using standard ASCII code requires $8 \times 16 = 128$ bits to store in a computer. A saving of almost 67%. Using the library (2.7), decoding the message (2.8) is straightforward since Huffman code is prefix-free, hence uniquely decodable. Note that the average codeword length of the code in (2.7) is given by:

$$\bar{l} = 4 \left(\frac{1}{16} \right) + 4 \left(\frac{1}{16} \right) + 3 \left(\frac{1}{8} \right) + 3 \left(\frac{1}{8} \right) + 3 \left(\frac{1}{8} \right) + 2 \left(\frac{1}{2} \right) + 2 \left(\frac{1}{2} \right) = \frac{21}{8}.$$

Note also that the entropy of the text message is:

$$H = -\sum_{i=1}^{7} p_i \log_2 (p_i) = - \left[\frac{1}{16} \log_2 \left(\frac{1}{16} \right) + \frac{1}{16} \log_2 \left(\frac{1}{16} \right) + \frac{1}{8} \log_2 \left(\frac{1}{8} \right) \right.$$

$$\left. + \frac{1}{8} \log_2 \left(\frac{1}{8} \right) + \frac{1}{8} \log_2 \left(\frac{1}{8} \right) + \frac{1}{4} \log_2 \left(\frac{1}{4} \right) + \frac{1}{4} \log_2 \left(\frac{1}{4} \right) \right] = \frac{21}{8}.$$

The fact that the average codeword length of the Huffman code is equal to the entropy of the source is expected in this example since each alphabet symbol appears with a probability equals to a negative power of 2.

2.9 Some remarks

We finish this chapter with some interesting remarks.

(1) *Huffman codes are not unique.* As seen in the above example, if three or more symbols have the same probability at any iteration, then the Huffman coding is not necessarily unique as it depends on the order in which these symbols are merged. While the decisions on how to merge equiprobable symbols may affect the individual codewords, they certainly have no effect on the average codeword length of the code. All Huffman codes will have the same (minimal) average codeword length.

(2) *There are optimal codes which are not Huffman codes.* Here is an example. Consider the alphabet $\mathcal{A} = \{\alpha_1, \ldots, \alpha_6\}$ with probability distribution vector $(0.08, 0.12, 0.15, 0.15, 0.24, 0.26)$. Applying the Huffman algorithm described above, it not difficult to see that the corresponding Huffman code $\mathcal{H}$ for the source is:

$$\alpha_1 \to 000, \; \alpha_2 \to 001, \; \alpha_3 \to 100, \; \alpha_4 \to 101, \; \alpha_5 \to 01, \; \alpha_6 \to 11. \tag{2.9}$$

Now, consider the following code $\mathcal{C}$ of the same alphabet:

$$\alpha_1 \to 000, \; \alpha_2 \to 100, \; \alpha_3 \to 001, \; \alpha_4 \to 101, \; \alpha_5 \to 01, \; \alpha_6 \to 11. \tag{2.10}$$

Clearly, $\mathcal{C}$ is optimal since it has the same average codeword length as the Huffman code $\mathcal{H}$. On the other hand, $\mathcal{C}$ is not a Huffman code on the alphabet $\mathcal{A}$ since the two least probable symbols (namely α_1 and α_2) do not have codewords that differ only in their last bit.

(3) The Huffman algorithm described above is called *static* because it assumes that the alphabet's probability distribution vector remains the same throughout the encoding (as well as the decoding) process. A large gap between an estimated source probability vector and the actual one can seriously deteriorate the efficiency of the Huffman code. A more flexible coding system is provided by the *Adaptive Huffman coding* where both the alphabet and the probabilities of its symbols are dynamic in the sense that they are updated frequently as the symbols enter the encoder.

2.10 References

Ida Mengyi Pu. (2006) *Fundamental Data Compression*, (Butterworth-Heinemann).

David Salomon. (2004). *Data Compression, The Complete Reference*, Third Edition, (Springer).

Chapter 3

The JPEG standard

3.1 Introduction

JPEG is an acronym for "Joint Picture Expert Group", the committee formed internationally in the 80's to create, develop and support global standards for compression of still (grayscale or colored) images. The committee was a result of a collaborative effort by three bodies: the International Telecommunication Union (ITU), the International Organization for Standardization (ISO) and the International Electrotechnical Commission (IEC). As noted on the official JPEG website (www.jpeg.org), people often use the term "JPEG" to refer to a particular compression standard and its implementation, not to the committee itself.

The JPEG standard defines four modes of operations in still image compression. We give a very brief description of each of these modes.

- **The Sequential lossy mode**. The image is broken into blocks. Each block is scanned once in a raster manner (left-to-right, top-to-bottom). Some information is lost during the compression and the reconstructed image is an approximation of the original one.
- **The Progressive mode**. Both compression and decompression of the image are done in several scans. Each scan produces a better image than the previous ones. The image is transferred starting with coarse resolution (almost unrecognisable) to finer resolution. In applications with long downloading time, the user will see the image building up in multiple resolutions.
- **The Hierarchical mode**. The image is encoded at multiple resolutions allowing applications to access a low resolution version without the need to decompress the full resolution version of the image. You

65

have probably noticed sometimes that you do not get the same quality when you print an image from the one displayed on a website since the two operations (printing and displaying) require different resolutions.

- **The Sequential lossless mode** The image is scanned once and encoded in a way that allows the exact recovery of every element of the image after decompression. This results, of course, in a much longer code stream than the ones obtained in the lossy modes.

The first three modes are called DCT-based modes since they all use the Discrete Cosine Transform (DCT for short, see Section 3.2) as the main tool to achieve compression. Each of the four modes has its own features and parameters that allow a certain degree of flexibility in terms of compression-to-quality ratio. The purpose of this chapter however is not to describe the technicalities and properties of the above modes nor to discuss the hardware implementation. We will be concerned only with the Sequential lossy mode implemented by the *Baseline* JPEG standard which can be described as the collection of "baseline routines" that must be included in every DCT-based JPEG standard. The Baseline standard is by far the most popular JPEG technique and it is well supported by almost all applications.

Although the Baseline standard applies to images with various color components, we will restrict our discussion to grayscale images only for simplicity. Once the technique is understood for grayscale images, it can be extended to color images with not much difficulty.

3.1.1 *Before you go further*

Mathematical skills required to have a good understanding of this chapter include basic matrix manipulations, basic linear algebra concepts like the notion of linear independence and basis. Also, some good knowledge of working with and simplifying trigonometric expressions is necessary.

3.2 The Discrete Cosine Transform (DCT)

The main ingredient in the JPEG Baseline compression recipe is a mathematical operation known as the Discrete Cosine Transform or DCT for short. In a nutshell, the DCT is a transformation that takes a signal data as an input and transforms it from one raw type of representation that usually contains an excess of information to another more suitable for applications. For example, if you think of a still image as a two-dimensional signal that is perceived by the human visual system, then the DCT can be used to convert the signal (or the spatial information) into a numeric data ("frequency" or "spectral" information) so that the image information exists in a quantitative form that can be manipulated for compression.

In this section, n is a fixed positive integer.

3.2.1   *The one-dimensional DCT*

Note first that there is more than one transform known as a discrete cosine transform in the literature. These transforms vary in minor details and are usually referred to as DCT-I to DCT-IV. The most popular one is the DCT-II known simply as the DCT. It is the transform used in the JPEG baseline standard and it is the one we consider in this chapter.

Given a list $\alpha = (a_0, a_1, \ldots, a_{n-1})$ of n real numbers, the one-dimensional discrete cosine transform of α is the list $\beta = (b_0, b_1, \ldots, b_{n-1})$ of n real numbers given by

$$b_j = \sum_{k=0}^{n-1} \sqrt{\frac{2}{n}} \gamma_j a_k \cos\left(\frac{(2k+1)j}{2n}\pi\right) \tag{3.1}$$

with

$$\gamma_j = \begin{cases} \frac{1}{\sqrt{2}} & \text{if } j = 0 \\ 1 & \text{if } j > 0. \end{cases} \tag{3.2}$$

Note in particular that $b_0 = \frac{a_0 + a_1 + \cdots + a_{n-1}}{\sqrt{n}}$ (since $\cos\left(\frac{j(2k+1)}{2n}\pi\right) = 1$ for all k when $j = 0$) is the mean value of the input list α. The coefficient b_0 is referred to as the DC coefficient, while the term AC coefficient is given to any b_j with $j > 0$.

This transformation has some interesting properties, but two properties in particular make it a valuable tool in data processing. The first one is its ability to concentrate most of the "energy" of a correlated sequence in a few transformed coefficients, usually the first ones. If the list α consists of correlated values, then very few coefficients in the transformed sequence β are large in absolute value and the rest are very small (close to zero in absolute value). The second important property of the DCT transformation is the fact that it is reversible. Given the transformed DCT coefficients $\beta = (b_0, b_1, \ldots, b_{n-1})$, then one can retrieve the original coefficients using the *inverse Discrete cosine transform* (IDCT for short) given by:

$$a_j = \sum_{k=0}^{n-1} \sqrt{\frac{2}{n}} \gamma_k b_k \cos\left(\frac{(2j+1)k}{2n}\pi\right) \qquad (3.3)$$

with γ_k like in (3.2). To put things in perspective, we look at an example. Consider the following (correlated) sequence of 8 terms

$$\alpha = (155, 150, 165, 154, 160, 167, 158, 163).$$

Applying the DCT to α leads to the following 8 coefficients (rounded to two decimal places):

$$\beta = (449.72, -8.39, -2.74, 0.10, -2.83, -0.99, 11.85, 3.55). \qquad (3.4)$$

Half of the transformed coefficients in β are smaller than 3 in absolute value. If we apply the IDCT to β, we get the following sequence

$$\alpha' = (155.0, 150.0, 164.9, 154.0, 160.0, 167.0, 158.0, 163.0),$$

which is almost the same as the original one, and this is what we expect from the inverse transform. What is less expected is the following fact: set 3 as a threshold point for the DCT coefficients in (3.4) in the sense that every coefficient less than 3 in absolute value is rounded to zero, and round the other coefficients to the nearest integer. We get the following new transformed sequence

$$\beta' = (450, -8, 0, 0, 0, 0, 12, 4).$$

When applying the IDCT again to β', we obtain the sequence

$$\alpha'' = (157.86, 149.11, 164.08, 154.06, 159.54, 165.20, 157.99, 164.92),$$

which is still relatively close to the original sequence. It is amazing how we can still almost reconstruct the original sequence with half of the eight DCT coefficients coarsely reduced to zero. It is precisely this kind of flexibility that makes the DCT such a useful tool for data compression. But what makes the DCT transformation behaves this way? To answer this question, we need to take a closer look at the operation of this transform. First consider the n vectors

$$\mathbf{d}_j = \sqrt{\frac{2}{n}} \gamma_j \left(\cos\left(j\left(0+\frac{1}{2}\right)\frac{\pi}{n}\right), \cos\left(j\left(1+\frac{1}{2}\right)\frac{\pi}{n}\right), \ldots, \cos\left(j\left(n-1+\frac{1}{2}\right)\frac{\pi}{n}\right)\right)$$

$$= \sqrt{\frac{2}{n}} \gamma_j \left(\cos\left(\frac{j\pi}{2n}\right), \cos\left(\frac{3j\pi}{2n}\right), \ldots, \cos\left(\frac{(2n-1)j\pi}{2n}\right)\right)$$

for $j = 0, 1, \ldots, n - 1$. Then, the dot product of the vector $\mathbf{d}_j$ with α (the original list) is

$$
\mathbf{d}_j \cdot \alpha = \sqrt{\frac{1}{n}} a_0 \cos\left(j\left(0 + \frac{1}{2}\right)\frac{\pi}{n}\right) + \sqrt{\frac{2}{n}} a_1 \cos\left(j\left(1 + \frac{1}{2}\right)\frac{\pi}{n}\right)
$$

$$
+ \sqrt{\frac{2}{n}} a_2 \cos\left(j\left(2 + \frac{1}{2}\right)\frac{\pi}{n}\right) + \cdots + \sqrt{\frac{2}{n}} a_{\text{n-1}} \cos\left(j\left(\text{n-1} + \frac{1}{2}\right)\frac{\pi}{n}\right)
$$

$$
= \sum_{k=0}^{n-1} \sqrt{\frac{2}{n}} \gamma_j a_k \cos\left(\frac{(2k+1)j}{2n}\pi\right) = b_j.
$$

Therefore, the DCT coefficient b_j is simply the dot product of the vector $\mathbf{d}_j$ with the original vector α. Next, consider the n cosine waves

$$
w_j(x) = \sqrt{\frac{2}{n}} \gamma_j \cos(jx), \quad j = 0, 1, \ldots, n - 1
$$

with the frequency of wave w_j being j. The vector $\mathbf{d}_j$ is formed by evaluating the wave w_j at each of the following n angles

$$
\frac{\pi}{2n}, \frac{3\pi}{2n}, \frac{5\pi}{2n}, \ldots, \frac{(2n-1)\pi}{2n}.
$$

For $n = 8$, the following table gives of the values of each of the eight waves $w_0, w_1, \ldots, w_7$ at each of the angles $\frac{\pi}{16}, \frac{3\pi}{16}, \frac{5\pi}{16}, \ldots, \frac{15\pi}{16}$.

x	$\pi/16$	$3\pi/16$	$5\pi/16$	$7\pi/16$	$9\pi/16$	$11\pi/16$	$13\pi/16$	$15\pi/16$
$w_0(x)$	0.3536	0.3536	0.3536	0.3536	0.3536	0.3536	0.3536	0.3536
$w_1(x)$	0.4904	0.4157	0.2778	0.0975	-0.0975	-0.2778	-0.4157	-0.4904
$w_2(x)$	0.4619	0.1913	-0.1913	-0.4619	-0.4619	-0.1913	0.1913	0.4619
$w_3(x)$	0.4157	0.0975	-0.4904	-0.2778	0.2778	0.4904	0.0975	-0.4157
$w_4(x)$	0.3536	-0.3536	-0.3536	0.3536	-0.3536	-0.3536	0.3536	0.3536
$w_5(x)$	0.2778	-0.4904	0.0975	0.4157	-0.4157	-0.0975	0.4904	-0.2778
$w_6(x)$	0.1913	-0.4619	0.4619	-0.1913	-0.1913	0.4619	-0.4619	0.1913
$w_7(x)$	0.0975	-0.2778	0.4157	-0.4904	0.4904	-0.4157	0.2778	-0.0975

The entries of the jth row in the table are the components of the vector $\mathbf{d}_j$ constructed above for $j = 0, 1, \ldots, 7$. The table shows that each vector $\mathbf{d}_j$, with the exception of vector $\mathbf{d}_0$, has four pairs of the form $(-\kappa, \kappa)$ for a certain coefficient κ. If the components of the list α are correlated, this results in a dot product $\mathbf{d}_j \cdot \alpha$ being relatively small in absolute value. Note also that the frequency j of the cosine wave $w_j(x)$ increases as we go down in the table. This means that the early DCT coefficients correspond to low-frequency components of the sequence and these components usually contain the important characteristics of the list. The table also reveals another important feature of the vectors $\mathbf{d}_j$: the dot product $\mathbf{d}_i \cdot \mathbf{d}_k$ is zero

The mathematics that power our world

for $i \neq k$ which means that the vectors are orthogonal.

The above table provides us as well with an efficient way to compute the transform vector β. Let D be the following 8×8 matrix:

$$
\begin{bmatrix}
0.3536 & 0.3536 & 0.3536 & 0.3536 & 0.3536 & 0.3536 & 0.3536 & 0.3536 \\
0.4904 & 0.4157 & 0.2778 & 0.0975 & -0.0975 & -0.2778 & -0.4157 & -0.4904 \\
0.4619 & 0.1913 & -0.1913 & -0.4619 & -0.4619 & -0.1913 & 0.1913 & 0.4619 \\
0.4157 & 0.0975 & -0.4904 & -0.2778 & 0.2778 & 0.4904 & 0.0975 & -0.4157 \\
0.3536 & -0.3536 & -0.3536 & 0.3536 & -0.3536 & -0.3536 & 0.3536 & 0.3536 \\
0.2778 & -0.4904 & 0.0975 & 0.4157 & -0.4157 & -0.0975 & 0.4904 & -0.2778 \\
0.1913 & -0.4619 & 0.4619 & -0.1913 & -0.1913 & 0.4619 & -0.4619 & 0.1913 \\
0.0975 & -0.2778 & 0.4157 & -0.4904 & 0.4904 & -0.4157 & 0.2778 & -0.0975
\end{bmatrix} \quad (3.5)
$$

then $\beta = D\alpha^t$ (matrix multiplication with α^t being the transpose of α).

In practical applications, the input sequence α is usually quite large. In this case, the sequence is broken into small subsequences containing n coefficients each. Each of the subsequences is treated separately and independently of the other segments in the source. Most applications using the DCT set 8 as the value of n.

3.2.2 *The two-dimensional DCT*

The one-dimensional DCT is used in practice in processing one-dimensional data such as speech signals. For the processing of two-dimensional data such as digital images, a two-dimensional version of the DCT is needed. In this case, the input data is represented by a square $n \times n$ matrix A (as opposed to a one-dimensional sequence α). The two-dimensional DCT transform of A is defined as being the matrix obtained from A by applying first the one-dimensional DCT to each column of A to get a matrix A_1, and then applying the one-dimensional DCT to each row of A_1. As we did for the one-dimensional DCT, we find a matrix expression for this transformation. First note that the matrix D given in (3.5) (found for $n = 8$) is a special case of the general matrix for arbitrary integer n known as the DCT matrix that we denote also by D and which is displayed in Figure 3.1.

Given an $n \times n$ matrix A, the columns of DA are the one-dimensional DCT transforms of the columns of A, and the rows of AD^t are the one-dimensional DCT transforms of the rows of A. So DAD^t is the transformation that applies the one-dimensional DCT first to the coulmns of A and then to the rows of the resulting matrix. Therefore, the two-dimensional

$$D = \begin{bmatrix} \sqrt{\frac{1}{n}} & \sqrt{\frac{1}{n}} & \cdots & \sqrt{\frac{1}{n}} \\ \sqrt{\frac{2}{n}}\cos\left(\frac{\pi}{2n}\right) & \sqrt{\frac{2}{n}}\cos\left(\frac{3\pi}{2n}\right) & \cdots & \sqrt{\frac{2}{n}}\cos\left(\frac{(2n-1)\pi}{2n}\right) \\ \sqrt{\frac{2}{n}}\cos\left(\frac{2\pi}{2n}\right) & \sqrt{\frac{2}{n}}\cos\left(\frac{6\pi}{2n}\right) & \cdots & \sqrt{\frac{2}{n}}\cos\left(\frac{2(2n-1)\pi}{2n}\right) \\ \vdots & \vdots & \cdots & \vdots \\ \sqrt{\frac{2}{n}}\cos\left(\frac{(n-1)\pi}{2n}\right) & \sqrt{\frac{2}{n}}\cos\left(\frac{3(n-1)\pi}{2n}\right) & \cdots & \sqrt{\frac{2}{n}}\cos\left(\frac{(n-1)(2n-1)\pi}{2n}\right) \end{bmatrix}$$

Fig. 3.1   Discrete Cosine Transform matrix.

DCT transform of A can be defined as being the $n \times n$ matrix B given by:

$$B = DAD^t \tag{3.6}$$

where D is the DCT matrix given in Figure 3.1. Relation (3.6) gives the following definition of the two-dimensional DCT transform in terms of the matrix coefficients. If $A = [a_{ij}]$, $0 \leq i, j \leq n - 1$ is the input matrix, then the two-dimensional DCT of A is the $n \times n$ matrix $B = [b_{ij}]$, $0 \leq i, j \leq n-1$ with the b_{ij} entry given by the expression

$$b_{ij} = \frac{2}{n}\gamma_i\gamma_j \sum_{k=0}^{n-1}\sum_{l=0}^{n-1} a_{kl} \cos\left(\frac{(2k+1)i\pi}{2n}\right)\cos\left(\frac{(2l+1)j\pi}{2n}\right) \tag{3.7}$$

and γ_r is defined as in (3.2). Similar to the one-dimensional DCT, the coefficient b_{00} in (3.7) is referred to as the *DC coefficient*, and every other coefficient b_{ij} is called an *AC coefficient*. The DC coefficient holds the mean value of all the coefficients in the matrix.

Like the one-dimensional DCT, the two-dimensional version is invertible. Given the matrix $B = [b_{ij}]$, $0 \leq i, j \leq n - 1$, then the inverse DCT transform of B is the matrix $A = [a_{ij}]$, $0 \leq i, j \leq n - 1$ given by:

$$a_{ij} = \frac{2}{n} \sum_{k=0}^{n-1}\sum_{l=0}^{n-1} \gamma_k\gamma_l b_{kl} \cos\left(\frac{(2i+1)k\pi}{2n}\right)\cos\left(\frac{(2j+1)l\pi}{2n}\right) \tag{3.8}$$

and γ_r is defined as in (3.2). From relation (3.6) above, we get that $A = D^{-1}B\left(D^t\right)^{-1}$, assuming D is an invertible matrix. We will prove in Section 3.5.2 that the DCT matrix is indeed invertible and satisfies a key property that makes it very useful in this application, namely $D^{-1} = D^t$. This shows that $A = D^t B D$.

3.3   DCT as a tool for image compression

We now come to the main application of this chapter. We explain how the Baseline JPEG standard uses the two-dimensional DCT to compress and decompress grayscale digital images. The procedure involves several steps that we explain and illustrate using the image in Example 3.1 below in each step.

3.3.1   *Image pre-processing*

At the encoder end, the source image is first divided into non-overlapping blocks of 8×8 pixels (called data units) from left to right, top to bottom. The size 8×8 for a data unit was chosen because at the time of development of the JPEG standard, that size fit well with the maximum size allowed by integrated circuit technology. Very often, the image is of size $m \times n$ pixels with either m or n (or both) not a multiple of 8. If m is not a multiple of 8, the bottom row is duplicated a number of times to get the nearest multiple of 8. If n is not a multiple of 8, the same thing is done on the rightmost column. For instance, for a 69×138 image, the last row is duplicated three times and the rightmost column is duplicated six times. Each pixel is represented by its *grayscale value* which indicates how bright that pixel is and each grayscale value is stored in a digital form as one *byte* (8 bits). Since $2^8 = 256$, each grayscale value is an integer in the range from 0 to 255 with 0 representing Black and 255 representing White. For example, a 69×138 picture requires (with duplications of the last row and the rightmost column) $72 \times 144 = 10368$ bytes to store before any compression is done. Each 8×8 block of the digital image is represented by an 8×8 matrix with integer entries in the range $[0, 255]$.

Example 3.1. Figure 3.2 is an 8×8 block taken somewhere from a digital image. The block is represented by the matrix A, which is an 8×8 matrix of the corresponding grayscale values. Note the $(4, 1)$-pixel (fourth row, first column) which seems to be the brightest in the block. The corresponding entry in A is the largest.

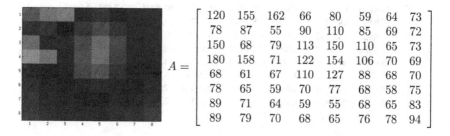

$$A = \begin{bmatrix} 120 & 155 & 162 & 66 & 80 & 59 & 64 & 73 \\ 78 & 87 & 55 & 90 & 110 & 85 & 69 & 72 \\ 150 & 68 & 79 & 113 & 150 & 110 & 65 & 73 \\ 180 & 158 & 71 & 122 & 154 & 106 & 70 & 69 \\ 68 & 61 & 67 & 110 & 127 & 88 & 68 & 70 \\ 78 & 65 & 59 & 70 & 77 & 68 & 58 & 75 \\ 89 & 71 & 64 & 59 & 55 & 68 & 65 & 83 \\ 89 & 79 & 70 & 68 & 65 & 76 & 78 & 94 \end{bmatrix}$$

Fig. 3.2   A raster image and its matrix.

3.3.2   *Level shifting*

As the values of the cosine function are centered at zero (the midpoint of the cosine range $[-1, 1]$), the DCT works more efficiently with values centered at 0 (negative and positive) rather than just positive integers. Since the midpoint of the range $[0, 255]$ is 127.5, we shift the grayscale values of the pixels from the interval $[0, 255]$ to $[-128, 127]$ by subtracting 128 from each entry in the input matrix. For the image in Example 3.1 above, the matrix of shifted values is

$$A_1 = \begin{bmatrix} -8 & 27 & 34 & -62 & -48 & -69 & -64 & -55 \\ -50 & -41 & -73 & -38 & -18 & -43 & -59 & -56 \\ 22 & -60 & -49 & -15 & 22 & -18 & -63 & -55 \\ 52 & 30 & -57 & -6 & 26 & -22 & -58 & -59 \\ -60 & -67 & -61 & -18 & -1 & -40 & -60 & -58 \\ -50 & -63 & -69 & -58 & -51 & -60 & -70 & -53 \\ -39 & -57 & -64 & -69 & -73 & -60 & -63 & -45 \\ -39 & -49 & -58 & -60 & -63 & -52 & -50 & -34 \end{bmatrix}.$$

3.3.3   *Applying the DCT*

After level-shifting, the two-dimensional DCT is applied independently to the shifted matrix A_1 of each block in the image. This is accomplished by computing $B = DA_1D^t$ where D is the DCT matrix with $n = 8$ given in (3.5). For the block in Example 3.1 above, the matrix B of DCT coefficients is the following.

$$B = \begin{bmatrix} -330.8750 & 65.1776 & -9.3885 & 46.0853 & 51.1250 & -30.8004 & -2.7408 & 4.0057 \\ 75.9481 & 58.8311 & -25.6708 & 8.7057 & -2.2776 & -20.4903 & -6.1304 & 18.0618 \\ -41.8828 & 3.8547 & 52.8126 & -60.6199 & -58.4390 & 0.5139 & 17.5914 & 12.1807 \\ -51.0028 & 1.0753 & 5.0003 & -54.8967 & -40.6112 & -6.6073 & 3.4048 & 11.2796 \\ 53.6250 & 41.9546 & 4.5161 & -16.0604 & -20.3750 & -14.3393 & -12.2887 & 5.2865 \\ 48.4935 & 67.3990 & 39.1726 & 6.2506 & -9.2437 & -6.9427 & 6.7856 & 6.1368 \\ 12.8836 & 14.1137 & 3.0914 & -3.2691 & 2.9643 & 7.5347 & 22.9374 & 16.8211 \\ -12.5815 & -23.0803 & -23.6630 & -11.5713 & 4.2610 & 17.0271 & 24.8242 & 16.5083 \end{bmatrix} \quad (3.9)$$

Instead of the 64 grayscale values, we now have an array of different spatial frequencies in the above block. It is probably worth mentioning at this point that if we apply the DCT transform directly to the original matrix A before the level shifting, we obtain the same matrix B with the exception of the DC coefficient (which would be 980.0000 instead of -44.0000). This has nothing to do with this particular example. It can be proven in general that shifting the coefficients of the matrix A by a constant value affects only the DC coefficient when the DCT is applied.

3.3.4 *Quantization*

So far no compression was made in the previous steps. Remember, the DCT is a completely invertible transformation. From the matrix B of DCT coefficients obtained in the previous step, we can recover the original matrix A of the unit data by computing $D^t B D$. Quantization is the step in the JPEG standard where the magic of compressing takes place. The information lost in this step is (in general) lost beyond recovery, and this is basically why we call the JPEG standard a "lossy" one. At this step, mathematics exploit the human eye perception and tolerance of what level of distortion is deemed acceptable to come up with a reasonable compression scheme. Experiments show that the human eye is more sensitive to the low frequency components of an image than the high frequency ones. The quantization step enables us to discard many of the high frequency coefficients as their presence have little effect on the perception of the image as a whole. After the DCT is applied to the 8×8 block, each of the 64 DCT coefficients of the block is first divided by the corresponding entry in the matrix $Q = [q_{ij}]$ below and then the result is rounded to the nearest integer:

$$Q = \begin{bmatrix} 16 & 11 & 10 & 16 & 24 & 40 & 51 & 61 \\ 12 & 12 & 14 & 19 & 26 & 58 & 60 & 55 \\ 14 & 13 & 16 & 24 & 40 & 57 & 69 & 56 \\ 14 & 17 & 22 & 29 & 51 & 87 & 80 & 62 \\ 18 & 22 & 37 & 56 & 68 & 109 & 103 & 77 \\ 24 & 35 & 55 & 64 & 81 & 104 & 113 & 92 \\ 49 & 64 & 78 & 87 & 103 & 121 & 120 & 101 \\ 72 & 92 & 95 & 98 & 112 & 100 & 103 & 99 \end{bmatrix}. \tag{3.10}$$

Designed based on human tolerance of visual effects, the matrix Q is called the *luminance quantization matrix*. It is not the only quantization matrix defined by the JPEG standard, but it is the one commonly used in applications. Note how the entries in the matrix Q increase almost in every row and every column as you move from left to right and top to bottom. This is designed to ensure aggressive quantization of coefficients with higher frequencies (the bigger the number we divide with, the closer the answer to zero). The quantization step has another important role to play in the JPEG standard. Notice how the coefficients of the DCT matrix B in (3.9) are real numbers rounded to four decimal places. When divided with the entries of Q, these DCT coefficients remain real-valued numbers and if we do not round them to create integer-valued integers, the encoding process of the quantized coefficients (see Section 3.3.5) would not be possible. The matrix B of DCT coefficients is transformed after quantization into the matrix $C = [c_{ij}]$ where $c_{ij} = Round\left(\frac{b_{ij}}{q_{ij}}\right)$. For the block in Example 3.1, the matrix C is the following.

$$C = \begin{bmatrix} -21 & 6 & -1 & 3 & 2 & -1 & 0 & 0 \\ 6 & 5 & -2 & 0 & 0 & 0 & 0 & 0 \\ -3 & 0 & 3 & -3 & -1 & 0 & 0 & 0 \\ -4 & 0 & 0 & -2 & -1 & 0 & 0 & 0 \\ 3 & 2 & 0 & 0 & 0 & 0 & 0 & 0 \\ 2 & 2 & 1 & 0 & 0 & 0 & 0 & 0 \\ 0 & 0 & 0 & 0 & 0 & 0 & 0 & 0 \\ 0 & 0 & 0 & 0 & 0 & 0 & 0 & 0 \end{bmatrix}. \tag{3.11}$$

3.3.5 Encoding

Now that we have the quantized matrix $C = [c_{ij}]$ from the last step, the next challenge is to encode the coefficients in a way to minimize the storage space. As it can be seen from (3.11), the quantization step is designed to create a matrix $C = [c_{ij}]$ with mostly zero coefficients c_{ij} for large values of i and j (in the lower right section of the matrix). To take advantage of the abundance of zeros in the matrix, the encoder starts by rearranging the quantized coefficients in one stream of 64 coefficients in a way to produce long runs of zeros. The idea is to assign a single code for a run of zeros rather than coding each zero individually. This can be best achieved by collecting the coefficients in a "zigzag" fashion starting with the DC coefficient in the upper left corner and ending with a sequence of trailing zeros with few short sequences of AC coefficients in between. The zigzag mode of collecting coefficients is illustrated in the following image:

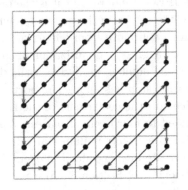

For the quantized matrix of Example 3.1, the zigzag mode yields the following string:

$$-21, 6, 6, -3, 5, -1, 3, -2, 0, -4, 3, 0, 3, 0, 2, -1, 0,$$
$$-3, 0, 2, 2, 0, 2, 0, -2, -1, 0, 0, 0, 0, 0, -1, 0, 1, EOB \qquad (3.12)$$

where the special symbol EOB (End Of Block) is introduced to indicate that all remaining coefficients after the last "1" are zero's till the end of the sequence. In our example, the last run consists of 30 consecutive zeros. As we will see later, the DC coefficient is encoded using a technique called differential encoding while the 63 AC coefficients are encoded using a run-length encoding technique. Huffman coding (see Chapter 2) is then used to encode both.

3.3.5.1 *Encoding the DC coefficients*

In a continuous tone picture, adjacent data units (8×8 pixel blocks) are usually closely correlated. Since the DC coefficient is proportional to the average of pixel intensities in the block, it is natural to assume that adjacent blocks in the image have fairly close DC coefficients. Once the DC coefficient DC_0 of the first block is encoded, it would make more sense to encode the difference $DC_1 - DC_0$ rather than encoding the DC coefficient DC_1 itself since the difference is small and would require fewer bits to encode. In general, the JPEG standard encodes the DC differences $d_i = DC_i - DC_{i-1}$ between DC coefficients of adjacent blocks rather than the usually large DC coefficients. For example, if the first four 8×8 blocks have respective (quantized) DC coefficients -21, -22, -18 and -18, then the JPEG standard encodes -21 in the first block, -1 in the second, 4 in the third and 0 in the fourth block (each of the codes is followed by the 63 codewords of the AC coefficients of the corresponding block). This technique is known as *differential encoding*.

For an image using 8 bits per pixel (that is what we assume for all what follows), it can be shown that both the AC and the DC coefficients fall in the range $[-1023, 1023]$ and the DC differences fall in the range $[-2047, 2047]$. The encoding of a DC difference coefficient d is done as follows:

(1) First find the minimal number γ of bits required to write $|d|$ (the absolute value of d) in binary form. For example, if $d = -7$, then $\gamma = 3$ since the binary form of $|-7|$ is 111. The minimal number of bits required to represent $|d|$ in binary form is referred to as the CATEGORY of d.

(2) Figure 3.3 shows the table used to encode the CATEGORY γ. This code is referred to as the *Variable-Length Code* of d or VLC for short. For example, the VLC for $d = -7$ is 100 (since $d = -7$ belongs to CATEGORY 3).

(3) The VLC code for d found in the previous step is the first layer in the encoding of d. The second layer is referred to as the *Variable-Length Integer* or VLI for short that we defined as follows. If $d \geq 0$, then VLI(d) consists of taking the γ least significant bits of the 8-bit binary representation of d (γ being, as above, the CATEGORY of d). If $d < 0$, then VLI(d) consists of taking the γ least significant bits of the 8-bit two's complement representation of $d - 1$ (see the first chapter on Calculator). Recall that the 8-bit two's complement representation of a negative integer β consists of writing the 8-bit binary representation of $|\beta|$, invert the digits (0 becomes 1, 1 becomes 0) and then add 1 to the answer. For example, if $\beta = -8$ then the 8-bit

binary representation of $|\beta|$ is 00001000 and its 8-bit two's complement representation is $11110111 + 1 = 11111000$. Since $d = -7$ belongs to CATEGORY 3, its VLI coding consists of taking the 3 least significant digits of the 8-bit two's complement representation of $-7 - 1 = -8$ which is 11111000. So VLI$(-7) = 000$.

(4) The final code of the DC difference d is VLC(d)VLI(d) (the concatenation of the Variable-Length Code and the Variable-Length Integer). For example, the codeword for $d = -7$ is 100000.

Category	$DC_i - DC_{i-1}$						Code
0	0						00
1	-1	1					010
2	-3	-2	2	3			011
3	-7	...	-4	4	...	7	100
4	-15	...	-8	8	...	15	101
5	-31	...	-16	16	...	31	110
6	-63	...	-32	32	...	63	1110
7	-127	...	-64	64	...	127	11110
8	-255	...	-128	128	...	255	111110
9	-511	...	-256	256	...	511	1111110
10	-1023	...	-512	512	...	1023	11111110
11	-2047	...	-1024	1024	...	2047	111111110

Fig. 3.3   Table for coding DC coefficients.

Assuming the figure given in Example 3.1 is the first block in a digital image, we encode its DC coefficient -21. The binary form of 21 is 10101, so -21 is Category 5. From the table in Figure 3.3, the VLC code of -21 is 110. The 8-bit two's complement of $-21 - 1 = -22$ is 11101010 and the 5 least significant bits of 11101010 are 01010. This is the VLI code of -21. We conclude that the encoding of -21 is 11001010.

3.3.5.2   *Encoding the AC coefficients*

The zigzag mode used to collect the quantized DCT coefficients is now put to work to encode the AC coefficients. Runs of zeros are efficiently compacted using a technique known as *run length encoding* (RLE for short). This technique is used to shorten a sequence containing runs of a repeated character by recording how many times the character appears in the run instead of actually listing it every single time. Huffman coding (see Chapter 2) of pairs of integers is combined with the RLE technique to produce a compressed binary sequence representing the AC coefficients in the 8×8 block. The sequence of the 63 AC coefficients is first shortened to a sequence of pairs and special symbols as follows: if α is a non-zero AC coefficient in the zigzag sequence, then α is replaced by the "object" $(r, m)(\alpha)$ where r is the length of the zero run immediately preceding α (that is the number of consecutive zeros preceding α), m is the CATEGORY of α from the above table. The maximum length of a run of zeros allowed in JPEG standard is 16. The symbol $(15, 0)$ is used to indicate a run of 16 zeros (one zero preceded by 15 other zeros). If a run has 17 zeros or more, it is divided into subruns of length 16 or less each. This means that r ranges between 0 and 15 and as a consequence it requires a 4-bit binary representation. In the intermediate sequence, the EOB is represented with the special pair $(0, 0)$. Let us illustrate using the AC coefficients in sequence (3.12) above resulting from the image in Example 3.1. The intermediate sequence for (3.12) is $(0, 3)(6)$; $(0, 3)(6)$; $(0, 2)(-3)$; $(0, 3)(5)$; $(0, 1)(-1)$; $(0, 2)(3)$; $(0, 2)(-2)$; $(1, 3)(-4)$; $(0, 2)(3)$; $(1, 2)(3)$; $(1, 2)(2)$; $(0, 1)(-1)$; $(1, 2)(-3)$; $(1, 2)(2)$; $(0, 2)(2)$; $(1, 2)(2)$; $(1, 2)(-2)$; $(0, 1)(-1)$; $(5, 1)(-1)$; $(1, 1)(1)$; $(0, 0)$. We explain how some of terms in this sequence are formed. The first non-zero AC coefficient is 6 with no preceding 0's. Since 6 belongs to CATEGORY 3, the first term in the intermediate sequence is $(0, 3)(6)$. The last AC coefficient -1 appearing in the zigzag sequence is preceded by a run of 5 zeros and since -1 belongs to CATEGORY 1, the corresponding entry in the intermediate sequence is $(5, 1)(-1)$.

Once the intermediate sequence is formed, each entry $(r, m)(\alpha)$ is encoded using the following steps.

(1) The pair (r, m) is encoded using Huffman codes provided by the standard tables in Section 3.3.5.3.
(2) The non-zero AC coefficient α is encoded using VLI codes as in the encoding of the DC difference coefficients above.
(3) The final codeword for $(r, m)(\alpha)$ is just the concatenation of codes of (r, m) and α.

The coding of the intermediate sequence of Example 3.1 is 100110; 100110; 0100; 100101; 000; 0111; 0101; 1111001011; 0111; 1101111; 1101110; 000;

1101100; 1101110; 0110; 1101110; 1101101; 000; 11110100; 11001; 1010.

Adding the code for the DC coefficient found earlier, the block image of Example 3.1 is encoded as:

11001010 100110 100110 0100 100101 000 0111 0101 1111001011 0111 1101111 1101110 000 1101100 1101110 0110 1101110 1101101 000 11110100 11001 1010.

Notice that the size of this new sequence is 124 bits. A saving of about 75% from the raw size of 512 ($8 \times 64 = 512$) bits if no compression was done on the block.

3.3.5.3   *AC coding tables*

The tables shown in this section are the recommended Huffman codes for AC (luminance) coefficients of the JPEG Baseline standard. They are based on statistics from experiments on a large number of images to classify "categories" of pixels values in terms of frequency of their occurrences in various images. As we saw in the chapter on Huffman coding, the idea is to assign shorter codewords for more frequent categories and longer codewords for less frequent ones. Note also that Huffman codes are uniquely decodable which leaves no room for ambiguity when the decompression process starts (see Section 3.4 below).

(Run, Category)	Codeword	Length
(0, 0)	1010	4
(0, 1)	00	2
(0, 2)	01	2
(0, 3)	100	3
(0, 4)	1011	4
(0, 5)	11010	5
(0, 6)	1111000	7
(0, 7)	11111000	8
(0, 8)	1111110110	10
(0, 9)	1111111110000010	16
(0, A)	1111111110000011	16

(Run, Category)	Codeword	Length
(1, 1)	1100	4
(1, 2)	11011	5
(1, 3)	1111001	7
(1, 4)	111110110	9
(1, 5)	11111110110	11
(1, 6)	1111111110000100	16
(1, 7)	1111111110000101	16
(1, 8)	1111111110000110	16
(1, 9)	1111111110000111	16
(1, A)	1111111110001000	16

(Run, Category)	Codeword	Length
(2, 1)	11100	5
(2, 2)	11111001	8
(2, 3)	1111110111	10
(2, 4)	111111110100	12
(2, 5)	1111111110001001	16
(2, 6)	1111111110001010	16
(2, 7)	1111111110001011	16
(2, 8)	1111111110001100	16
(2, 9)	1111111110001101	16
(2, A)	1111111110001110	16

(Run, Category)	Codeword	Length
(3, 1)	111010	6
(3, 2)	111110111	9
(3, 3)	111111110101	12
(3, 4)	1111111110001111	16
(3, 5)	1111111110010000	16
(3, 6)	1111111110010001	16
(3, 7)	1111111110010010	16
(3, 8)	1111111110010011	16
(3, 9)	1111111110010100	16
(3, A)	1111111110010101	16

(Run, Category)	Codeword	Length
(4, 1)	111011	6
(4, 2)	1111111000	10
(4, 3)	1111111110010110	16
(4, 4)	1111111110010111	16
(4, 5)	1111111110011000	16
(4, 6)	1111111110011001	16
(4, 7)	1111111110011010	16
(4, 8)	1111111110011011	16
(4, 9)	1111111110011100	16
(4, A)	1111111110011101	16

(Run, Category)	Codeword	Length
(5, 1)	1111010	7
(5, 2)	11111110111	11
(5, 3)	1111111110011110	16
(5, 4)	1111111110011111	16
(5, 5)	1111111110100000	16
(5, 6)	1111111110100001	16
(5, 7)	1111111110100010	16
(5, 8)	1111111110100011	16
(5, 9)	1111111110100100	16
(5, A)	1111111110100101	16

(Run, Category)	Codeword	Length
(6, 1)	1111011	7
(6, 2)	111111110110	12
(6, 3)	1111111110100110	16
(6, 4)	1111111110100111	16
(6, 5)	1111111110101000	16
(6, 6)	1111111110101001	16
(6, 7)	1111111110101010	16
(6, 8)	1111111110101011	16
(6, 9)	1111111110101100	16
(6, A)	1111111110101101	16

(Run, Category)	Codeword	Length
(7, 1)	11111010	8
(7, 2)	111111110111	12
(7, 3)	1111111110101110	16
(7, 4)	1111111110101111	16
(7, 5)	1111111110110000	16
(7, 6)	1111111110110001	16
(7, 7)	1111111110110010	16
(7, 8)	1111111110110011	16
(7, 9)	1111111110110100	16
(7, A)	1111111110110101	16

(Run, Category)	Codeword	Length
(8, 1)	111111000	9
(8, 2)	111111111000000	15
(8, 3)	1111111110110110	16
(8, 4)	1111111110110111	16
(8, 5)	1111111110111000	16
(8, 6)	1111111110111001	16
(8, 7)	1111111110111010	16
(8, 8)	1111111110111011	16
(8, 9)	1111111110111100	16
(8, A)	1111111110111101	16

(Run, Category)	Codeword	Length
(9, 1)	111111001	9
(9, 2)	111111110111110	16
(9, 3)	1111111110111111	16
(9, 4)	1111111111000000	16
(9, 5)	1111111111000001	16
(6, 6)	1111111111000010	16
(9, 7)	1111111111000011	16
(9, 8)	1111111111000100	16
(9, 9)	1111111111000101	16
(9, A)	1111111111000110	16

(Run, Category)	Codeword	Length
(A, 1)	111111010	9
(A, 2)	1111111111000111	16
(A, 3)	1111111111001000	16
(A, 4)	1111111111001001	16
(A, 5)	1111111111001010	16
(A, 6)	1111111111001011	16
(A, 7)	1111111111001100	16
(A, 8)	1111111111001101	16
(A, 9)	1111111111001110	16
(A, A)	1111111111001111	16

(Run, Category)	Codeword	Length
(B, 1)	1111111001	10
(B, 2)	1111111111010000	16
(B, 3)	1111111111010001	16
(B, 4)	1111111111010010	16
(B, 5)	1111111111010011	16
(B, 6)	1111111111010100	16
(B, 7)	1111111111010101	16
(B, 8)	1111111111010110	16
(B, 9)	1111111111010111	16
(B, A)	1111111111011000	16

(Run, Category)	Codeword	Length
(C, 1)	1111111010	10
(C, 2)	1111111111011001	16
(C, 3)	1111111111011010	16
(C, 4)	1111111111011011	16
(C, 5)	1111111111011100	16
(C, 6)	1111111111011101	16
(C, 7)	1111111111011110	16
(C, 8)	1111111111011111	16
(C, 9)	1111111111100000	16
(C, A)	1111111111100001	16

(Run, Category)	Codeword	Length
(D, 1)	11111111000	11
(D, 2)	1111111111100010	16
(D, 3)	1111111111100011	16
(D, 4)	1111111111100100	16
(D, 5)	1111111111100101	16
(D, 6)	1111111111100110	16
(D, 7)	1111111111100111	16
(D, 8)	1111111111101000	16
(D, 9)	1111111111101001	16
(D, A)	1111111111101010	16

(Run, Category)	Codeword	Length
(E, 1)	1111111111101011	16
(E, 2)	1111111111101100	16
(E, 3)	1111111111101101	16
(E, 4)	1111111111101110	16
(E, 5)	1111111111101111	16
(E, 6)	1111111111110000	16
(E, 7)	1111111111110001	16
(E, 8)	1111111111110010	16
(E, 9)	1111111111110011	16
(E, A)	1111111111110100	16

(Run, Category)	Codeword	Length
(F, 0)(ZRL)	11111111001	11
(F, 1)	1111111111110101	16
(F, 2)	1111111111110110	16
(F, 3)	1111111111110111	16
(F, 4)	1111111111111000	16
(F, 5)	1111111111111001	16
(F, 6)	1111111111111010	16
(F, 7)	1111111111111011	16
(F, 8)	1111111111111100	16
(F, 9)	1111111111111101	16
(F, A)	1111111111111110	16

3.4 JPEG decompression

The decompression aims to reconstruct the image from the compressed stream. This is achieved by reversing the above steps in the compression process. The decompression is done independently on each block and at the end of the process, all the decompressed blocks are merged together to form the reconstructed image.

(1) When the compressed binary stream of the block enters the decoder gate, it is read bit by bit. Using the Huffman encoding tables given in the previous section, the decoder reconstructs the intermediate sequence of objects $(r, m)(\alpha)$.

(2) From this intermediate sequence, the quantized DC coefficient, all the 63 quantized AC coefficients and all the run lengths can be reconstructed in same zigzag ordering as in the encoding step above. Recall that the first part of the compressed stream represents the (quantized) DC difference coefficient $d_i = DC_i - DC_{i-1}$. The quantized DC coefficient of block i is reconstructed as $DC_i = DC_{i-1} + d_i$ for $i \geq 1$ (assuming the DC coefficient D_{i-1} of block $i - 1$ was obtained at the previous step).

(3) The sequence obtained in the previous step is "dezigzaged" to reconstruct the 8×8 matrix of the quantized DCT coefficients of the block. For the block in Example 3.1, this step will reproduce the matrix (3.11) above.

(4) The 8×8 matrix of the quantized DCT coefficients is "dequantized" by multiplying each of the entries with the corresponding entry of the quantization matrix Q given in (3.10) above. For the block in Example 3.1, this step will produce the following matrix.

$$S = \begin{bmatrix} -336 & 66 & -10 & 48 & 48 & -40 & 0 & 0 \\ 72 & 60 & -28 & 0 & 0 & 0 & 0 & 0 \\ -42 & 0 & 48 & -72 & -40 & 0 & 0 & 0 \\ -56 & 0 & 0 & -58 & -51 & 0 & 0 & 0 \\ 54 & 44 & 0 & 0 & 0 & 0 & 0 & 0 \\ 48 & 70 & 55 & 0 & 0 & 0 & 0 & 0 \\ 0 & 0 & 0 & 0 & 0 & 0 & 0 & 0 \\ 0 & 0 & 0 & 0 & 0 & 0 & 0 & 0 \end{bmatrix}.$$

Notice that the matrix S is close to the original matrix B of the DCT coefficient given in (3.9) above but not exactly the same. This is due to the fact that entries in B were rounded to the nearest integers after quantization is performed.

(5) Now apply the two-dimensional inverse DCT given in (3.8) to the matrix S to get the matrix B_1.

$$
B_1 = \begin{bmatrix}
-9.5363 & 26.1090 & 13.8352 & -36.5921 & -64.4003 & -71.7877 & -68.6127 & -54.1596 \\
-58.6093 & -53.3551 & -52.3111 & -35.6142 & -18.6246 & -39.9788 & -64.1639 & -56.5378 \\
19.1002 & -33.7537 & -63.8441 & -22.2048 & 16.3370 & -20.4071 & -62.0797 & -52.9300 \\
61.4794 & -0.7968 & -43.2273 & -10.3405 & 23.8693 & -16.4274 & -63.1244 & -58.2510 \\
-44.8950 & -54.4807 & -59.1986 & -29.7012 & 0.2449 & -25.1636 & -64.9681 & -69.2678 \\
-79.9202 & -66.0041 & -69.1116 & -67.7725 & -51.6672 & -55.3606 & -66.3567 & -58.5642 \\
-30.3389 & -34.5392 & -61.1915 & -78.2150 & -71.1730 & -70.3833 & -63.1596 & -37.2653 \\
-41.7867 & -52.7451 & -69.2971 & -61.0331 & -42.9813 & -53.3216 & -56.7931 & -30.6488
\end{bmatrix}.
$$

(6) Round the entries of B_1 to the nearest integer:

$$
B_2 = \begin{bmatrix}
-10 & 26 & 14 & -37 & -64 & -72 & -69 & -54 \\
-59 & -53 & -52 & -36 & -19 & -40 & -64 & -57 \\
19 & -34 & -64 & -22 & 16 & -20 & -62 & -53 \\
61 & -1 & -43 & -10 & 24 & -16 & -63 & -58 \\
-45 & -54 & -59 & -30 & 0 & -25 & -65 & -69 \\
-80 & -66 & -69 & -68 & -52 & -55 & -66 & -59 \\
-30 & -35 & -61 & -78 & -71 & -70 & -63 & -37 \\
-42 & -53 & -69 & -61 & -43 & -53 & -57 & -31
\end{bmatrix}.
$$

(7) Add 128 to each entry of B_2:

$$
A' = \begin{bmatrix}
118 & 154 & 142 & 91 & 64 & 56 & 59 & 74 \\
69 & 75 & 76 & 92 & 109 & 88 & 64 & 71 \\
147 & 94 & 64 & 106 & 144 & 108 & 66 & 75 \\
189 & 127 & 85 & 118 & 152 & 112 & 65 & 70 \\
83 & 74 & 69 & 98 & 128 & 103 & 63 & 59 \\
48 & 62 & 59 & 60 & 76 & 73 & 62 & 69 \\
98 & 93 & 67 & 50 & 57 & 58 & 65 & 91 \\
86 & 75 & 59 & 67 & 85 & 75 & 71 & 97
\end{bmatrix}.
$$

(8) Matrix A' represents the grayscale values of the reconstructed 8×8 block.

The original and the reconstructed images are shown:

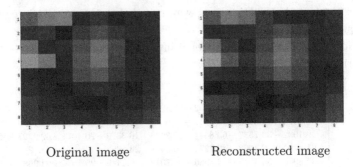

Original image   Reconstructed image

Not bad, considering that the compressed image takes only about 25% of the storage space taken by the raw image.

3.5   The mathematics of DCT

In this section, we dig deeper into the mathematics that make the DCT such an important tool for image compression. As the reader will soon discover, all the magic of JPEG compression happens by mixing together some basic properties of linear algebra with some trigonometric identities.

3.5.1   *Two-dimensional DCT as a linear tranformation*

From the perspective of relation (3.6) above, the two-dimensional DCT can be seen as a transformation:

$$\phi : \mathbb{M}_n \to \mathbb{M}_n$$
$$A \mapsto DAD^t \tag{3.13}$$

where $\mathbb{M}_n$ is the vector space of all $n \times n$ real matrices and D as in Figure 3.1 on page 71. The transformation ϕ is clearly linear:

$$\phi\left(\alpha A_1 + \beta A_2\right) = D\left(\alpha A_1 + \beta A_2\right) D^t = \alpha D A_1 D^t + \beta D A_2 D^t$$
$$= \phi\left(A_1\right) + \phi\left(A_2\right)$$

for any $A_1, A_2 \in \mathbb{M}_n$ and $\alpha, \beta \in \mathbb{R}$.

The key property of the DCT matrix that makes it attractive to real life applications is the fact that it is an *orthogonal* matrix. Before we proceed to the definition of orthogonality, let us quickly review some notions from linear algebra.

Consider a set of nonzero vectors $\Sigma = \{v_1, \ldots, v_s\}$ of $\mathbb{R}^r$ (where r is a positive integer).

- Σ is called *linearly independent* if the only way to have an equation of the form

$$a_1 v_1 + a_2 v_2 + \cdots + a_s v_s = 0$$

with $a_k \in \mathbb{R}$ for all k is that $a_1 = a_2 = \cdots = a_s = 0$.
- Σ is called a *spanning set* of $\mathbb{R}^r$ if every vector $v \in \mathbb{R}^r$ can be written as a linear combination of the vectors in Σ:

$$v = a_1 v_1 + a_2 v_2 + \cdots + a_s v_s, \quad a_k \in \mathbb{R} \text{ for all } k.$$

- Σ is called a *basis* of $\mathbb{R}^r$ if it is at the same time a spanning set and linearly independent. For example, the set

$$\{(1, 0, \ldots, 0), (0, 1, 0, \ldots, 0), \ldots, (0, 0, \ldots, 0, 1)\}$$

is a basis of $\mathbb{R}^r$ called the *standard basis*. Note that in $\mathbb{R}^r$, any basis contains exactly r (linearly independent) vectors.
- if $x = (x_1, \ldots, x_r)$ and $y = (y_1, \ldots, y_r)$ are two vectors in $\mathbb{R}^r$, then their *dot product* is $x \cdot y = x_1 y_1 + \cdots + x_r y_y$. In particular $x \cdot x = \sum_{i=1}^{r} x_i^2$ is equal to $\|x\|^2$ where $\|x\|$ is the magnitude (or the norm) of the vector x. In the case where $x \cdot x = 1$, x is called a *unit vector*. If x and y are nonzero vectors such that $x \cdot y = 0$, we say that they are *orthogonal*.
- The set Σ is called *orthogonal* if

$$v_i \cdot v_j = 0, \quad \text{for all } i \neq j, \ i, j \in \{1, 2, \ldots, s\}.$$

If in addition, every vector in Σ is a unit vector, the set Σ is called *orthonormal*.

- A basis $\mathcal{B}$ is called an *orthonormal basis* of $\mathbb{R}^r$ if it is an orthonormal set in addition of being a basis. For example, the standard basis of $\mathbb{R}^r$ is an orthonormal basis.

Definition 3.1. A matrix A is called orthogonal if

$$AA^t = A^t A = I \tag{3.14}$$

where A^t denotes, as usual, the transpose of the matrix A and I is the identity matrix (square matrix with all 1's on the main diagonal and 0's elsewhere).

Note that relation (3.14) implies in particular that A is a square invertible matrix and that

$$A^{-1} = A^t. \tag{3.15}$$

From a compression point of view, relation (3.15) is extremely valuable since it provides a clean and a "low cost" way of reversing the compression procedure and of reconstructing the original data. The relation also allows us to look at the DCT matrix as a "transition matrix" (change of basis matrix) from the standard basis E_{ij}, $i, j = 0, \ldots, n - 1$ (the entries of E_{ij} are all zeros except for the one on the ith row and the jth column which is 1) of $\mathbb{M}_n$ to another basis S much more useful to the compression standard. To see what the elements of the basis S look like, we go back to the definition of the two-dimensional DCT and its inverse. Given an $n \times n$ matrix $A = [a_{ij}]$, its DCT transform is given by the matrix $B = [b_{ij}]$ with $B = DAD^t$ and D is the DCT matrix displayed in Figure 3.1 on page 71. The inverse DCT is given by $A = D^t BD$ as explained above. Write $B = \sum_{i=0}^{n-1} \sum_{j=0}^{n-1} b_{ij} E_{ij}$, then

$$A = D^t BD = D^t \left(\sum_{i=0}^{n-1} \sum_{j=0}^{n-1} b_{ij} E_{ij} \right) D = \sum_{i=0}^{n-1} \sum_{j=0}^{n-1} b_{ij} \left(D^t E_{ij} D \right).$$

Let $S_{ij} = D^t E_{ij} D$ for $i \neq j$. So the matrix A can be expressed as $A = \sum b_{ij} S_{ij}$ and the matrices S_{ij} are precisely the elements of the new basis S of $\mathbb{M}_n$ in which every matrix A can be expressed as a linear combination of the matrices S_{ij} with coefficients equal to the DCT coefficients b_{ij} of A. Clearly, the coefficients of S_{ij} are not in the range $[0, 255]$ which makes them unsuitable to represent grayscale images. To see what the blocks S_{ij}'s look like as grayscale images, we proceed as follows.

(1) First notice that the (k, l)-coefficient of the matrix S_{ij} is given by

$$s_{kl} = \frac{\delta_i \delta_j}{n} \cos \left(\frac{i(2k + 1)\pi}{2n} \right) \cos \left(\frac{j(2l + 1)\pi}{2n} \right)$$

with $\delta_r = \sqrt{2}$ for $r > 0$ and $\delta_0 = 1$. So, multiplying the matrix S_{ij} by the factor $\frac{n}{\delta_i \delta_j}$ produces a matrix S'_{ij} with coefficients in the range $[-1, 1]$.

(2) If $-1 \leq x \leq 1$, then $0 \leq x + 1 \leq 2$ and $0 \leq \frac{255}{2}(x + 1) \leq 255$. So adding 1 to each coefficient of S'_{ij} and then scaling the coefficients by a factor of $\frac{255}{2}$ will produce a matrix S''_{ij} with coefficients in the range $[0, 255]$ and hence can be interpreted as a grayscale image.

The following picture gives a representation of each of the elements of the new basis $\mathcal{S}$ as a grayscale image. Note that the upper left corner of the array is just a white image resulting from the fact that each coefficient in the transformed matrix S''_{00} is equal to 255 (pure white).

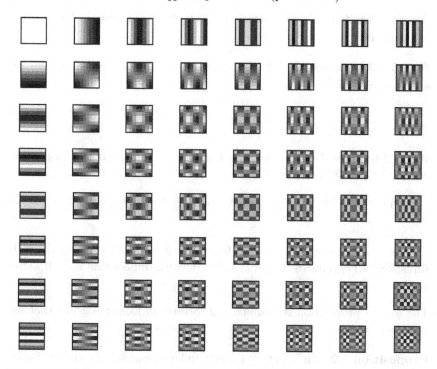

The fact that the coefficients of the matrix A in the new basis $\mathcal{S}$ are precisely the DCT coefficients of A allows us to interpret these coefficients as a "measure" of how much each of the squares in the above array is present in the image.

3.5.2 What is the deal with orthogonal bases anyway?

You might be wondering at this point why the name "orthogonal" for an $n \times n$ matrix A satisfying $AA^t = I$? The following theorem gives the answer.

Theorem 3.1. For an $n \times n$ matrix A, the following conditions are equivalent.

(1) A is an orthogonal matrix.

(2) The columns of A form an orthonormal basis of $\mathbb{R}^n$.
(3) The rows of A form an orthonormal basis of $\mathbb{R}^n$.

For the proof, we need the following two propositions.

Proposition 3.1. Every orthogonal set of $\mathbb{R}^r$ is in particular linearly independent.

Proof. Assume that $\Sigma = \{v_1, \ldots, v_s\}$ is an orthogonal set of $\mathbb{R}^r$ and let $a_1, \ldots, a_s \in \mathbb{R}$ such that

$$\sum_{k=1}^{s} a_k v_k = 0. \tag{3.16}$$

We need to prove that $a_k = 0$ for all $k = 1, 2, \ldots, s$. Multiplying (using dot product) equation (3.16) by v_k, we get:

$$a_1(v_1 \cdot v_k) + a_2(v_2 \cdot v_k) + \cdots + a_k(v_k \cdot v_k) + \cdots + a_s(v_s \cdot v_k) = 0. \tag{3.17}$$

Since Σ is orthogonal, $v_i \cdot v_k = 0$ for all $i \neq k$ and $v_k \cdot v_k = \|v_k\|^2 \neq 0$. Equation (3.17) reduces to $a_k(v_k \cdot v_k) = 0$ which implies that $a_k = 0$. This is true for any $k = 1, 2, \ldots, s$. □

The second proposition is a fundamental result in Linear Algebra that we state here without proof.

Proposition 3.2. In $\mathbb{R}^r$, every linearly independent set of r vectors forms a basis of $\mathbb{R}^r$.

For instance, to prove that the set $\left\{ q_1 = \begin{bmatrix} 1 \\ -1 \end{bmatrix}, \ q_2 = \begin{bmatrix} 1 \\ 0 \end{bmatrix} \right\}$ forms a basis of $\mathbb{R}^2$, it is enough to show that they are linearly independent. It is clearly the case since none of these vectors is a scalar multiple of the other.

Combining the above propositions, we get the following.

Corollary 3.1. An orthonormal set of r vectors in $\mathbb{R}^r$ forms an orthonormal basis of $\mathbb{R}^r$.

We proceed now to prove Theorem 3.1. Note first that since the statement "A is orthogonal" is equivalent to "A^t is orthogonal", we only need to prove that the first two conditions of Theorem 3.1 are equivalent. Write

$A = \begin{bmatrix} q_1 & q_2 & \cdots & q_n \end{bmatrix}$ where q_i is the ith column of A. Then $A^t = \begin{bmatrix} q_1^t \\ q_2^t \\ \vdots \\ q_n^t \end{bmatrix}$ and

therefore

$$A^t A = \begin{bmatrix} q_1^t \cdot q_1 & q_1^t \cdot q_2 & \cdots & q_1^t \cdot q_n \\ q_2^t \cdot q_1 & q_2^t \cdot q_2 & \cdots & q_2^t \cdot q_n \\ \vdots & \vdots & \cdots & \vdots \\ q_n^t \cdot q_1 & q_n^t \cdot q_2 & \cdots & q_n^t \cdot q_n \end{bmatrix} \tag{3.18}$$

where, as usual, $q_i^t \cdot q_j$ is the dot product of the vectors q_i^t and q_j. If A is an orthonormal matrix, then $A^t A = I$ and by comparing (3.18) with the identity matrix I, we conclude that $q_i^t \cdot q_j = 0$ for all i, j with $i \neq j$ and $q_i^t \cdot q_i = 1$ for all i. These relations show that the columns of A are orthogonal and unit vectors of $\mathbb{R}^n$. Corollary 3.1 implies that the columns of A form an orthonormal basis of $\mathbb{R}^n$. This proves the implication $(1) \Rightarrow (2)$ of Theorem 3.1. The implication $(2) \Rightarrow (1)$ is proven similarly.

3.5.3 *Proof of the orthogonality of the DCT matrix*

In this section, we give a detailed proof of the following result using Theorem 3.1.

Theorem 3.2. The DCT matrix defined in Figure 3.1 on page 71 is orthogonal.

There are few elegant and relatively short proofs of this result in the literature, but they require some heavy mathematics. The proof presented here is technical and somewhat long, but stays close to the basics. Let us start by recalling some trigonometric identities needed for the proof.

$$\cos(\alpha) \cos(\beta) = \frac{1}{2} \left[\cos(\alpha + \beta) + \cos(\alpha - \beta) \right]. \tag{3.19}$$

$$\cos^2(\alpha) = \frac{1}{2} \left[1 + \cos(2\alpha) \right]. \tag{3.20}$$

Lemma 3.1. For $j \in \{1, 2, \ldots, 2n - 1\}$:

$$\sum_{m=0}^{2n-1} \cos\left(\frac{mj\pi}{n} \right) = 0.$$

Proof. First recall that if $x \in \mathbb{R}$, then the *complex exponential* e^{ix} is defined as follows

$$e^{ix} = \cos x + i \sin x$$

where i is the complex number satisfying $i^2 = -1$. Fix $j \in \{1, 2, \ldots, 2n-1\}$ and let $\alpha = \frac{j\pi}{2n}$. Then:

$$\sum_{m=0}^{2n-1} \left(e^{2i\alpha}\right)^m = \sum_{m=0}^{2n-1} e^{2mi\alpha} = 1 + e^{2i\alpha} + e^{4i\alpha} + \cdots + e^{2(2n-1)i\alpha}.$$

The expression $1 + e^{2i\alpha} + e^{4i\alpha} + \cdots + e^{2(2n-1)i\alpha}$ is a geometric sum containing $2n$ terms with 1 as the first term and the ratio of any two consecutive terms equals to $e^{2i\alpha}$. Note that $0 < 2\alpha < 2\pi$ since $2\alpha = \frac{j}{n}\pi$ and $1 \leq j \leq 2n - 1$. Therefore the equations $\cos(2\alpha) = 1$ and $\sin(2\alpha) = 0$ cannot be satisfied simultaneously. Since $e^{2i\alpha} = \cos(2\alpha) + i\sin(2\alpha) = 1$ if and only if $\cos(2\alpha) = 1$ and $\sin(2\alpha) = 0$, we can confirm that the ratio $e^{2i\alpha}$ of the above geometric sum is not 1. A well-known formula for a finite geometric sum (with ratio other than 1) allows us to write:

$$\sum_{m=0}^{2n-1} \left(e^{2i\alpha}\right)^m = \frac{1 - \left(e^{2i\alpha}\right)^{2n}}{1 - e^{2i\alpha}} = \frac{1 - e^{4ni\alpha}}{1 - e^{2i\alpha}}. \tag{3.21}$$

Note that $e^{4ni\alpha} = \cos(4n\alpha) + i\sin(4n\alpha) = \underbrace{\cos(2j\pi)}_{1} + i\underbrace{\sin(2j\pi)}_{0} = 1$, which implies (by equation (3.21)) that $\sum_{m=0}^{2n-1} \left(e^{2i\alpha}\right)^m = 0$. We conclude that

$$0 = \sum_{m=0}^{2n-1} e^{2mi\alpha} = \sum_{m=0}^{2n-1} \left(\cos(2m\alpha) + i\sin(2m\alpha)\right)$$

$$= \sum_{m=0}^{2n-1} \left(\cos\left(\frac{mj\pi}{n}\right) + i\sin\left(\frac{mj\pi}{n}\right)\right)$$

$$= \sum_{m=0}^{2n-1} \cos\left(\frac{mj\pi}{n}\right) + i\sum_{m=0}^{2n-1} \sin\left(\frac{mj\pi}{n}\right).$$

Since both real and imaginary parts of the last complex expression must be zero. The result of the lemma follows. $\square$

Lemma 3.2. If $j \in \{1, 2, \ldots, 2n - 1\}$ is **even**, then:

$$\sum_{m=0}^{n-1} \cos\left(\frac{mj\pi}{n}\right) = 0.$$

Proof. Write $j = 2\alpha$ for some $\alpha \in \mathbb{Z}$. Using the result of Lemma 3.2, we have:

$$0 = \sum_{m=0}^{2n-1} \cos\left(\frac{mj\pi}{n}\right) = \sum_{m=0}^{n-1} \cos\left(\frac{mj\pi}{n}\right) + \sum_{m=0}^{n-1} \cos\left(\frac{(m+n)j\pi}{n}\right)$$

$$= \sum_{m=0}^{n-1} \cos\left(\frac{mj\pi}{n}\right) + \sum_{m=0}^{n-1} \cos\left(\frac{2\alpha m\pi}{n} + 2\alpha\pi\right)$$

$$= \sum_{m=0}^{n-1} \cos\left(\frac{mj\pi}{n}\right) + \sum_{m=0}^{n-1} \cos\left(\frac{\overbrace{2\alpha}^{j} m\pi}{n}\right)$$

since $\cos(y + 2\alpha\pi) = \cos(y)$. Thus,

$$0 = \sum_{m=0}^{2n-1} \cos\left(\frac{mj\pi}{n}\right) = 2\sum_{m=0}^{n-1} \cos\left(\frac{mj\pi}{n}\right)$$

and consequently,

$$\sum_{m=0}^{n-1} \cos\left(\frac{mj\pi}{n}\right) = 0$$

for $j \in \{1, 2, \ldots, 2n-1\}$ even. $\qquad\square$

Lemma 3.3. For $j \in \{1, 2, \ldots, 2n-1\}$ **odd:**

$$\sum_{m=1}^{n-1} \cos\left(\frac{mj\pi}{n}\right) = 0.$$

Proof. The proof of this lemma is a bit more technical and requires dividing it into subcases. First assume that n is even. Then

$$\sum_{m=1}^{n-1} \cos\left(\frac{mj\pi}{n}\right) = \sum_{m=1}^{\frac{n}{2}-1} \cos\left(\frac{mj\pi}{n}\right) + \cos\left(\frac{\frac{n}{2}j\pi}{n}\right) + \sum_{m=\frac{n}{2}+1}^{n-1} \cos\left(\frac{mj\pi}{n}\right).$$

Note that $\cos\left(\frac{\frac{n}{2}j\pi}{n}\right) = \cos\left(j\frac{\pi}{2}\right) = 0$ since j is odd (the cosine of any odd multiple of $\frac{\pi}{2}$ is zero). So

$$\sum_{m=1}^{n-1} \cos\left(\frac{mj\pi}{n}\right) = \sum_{m=1}^{\frac{n}{2}-1} \cos\left(\frac{mj\pi}{n}\right) + \sum_{m=\frac{n}{2}+1}^{n-1} \cos\left(\frac{mj\pi}{n}\right). \tag{3.22}$$

Now, let $k = n - m$, then the last sum in (3.22) can be written as:

$$\sum_{m=\frac{n}{2}+1}^{n-1} \cos\left(\frac{mj\pi}{n}\right) = \sum_{k=1}^{\frac{n}{2}-1} \cos\left(\frac{(n-k)j\pi}{n}\right) = \sum_{k=1}^{\frac{n}{2}-1} \cos\left(j\pi - \frac{kj\pi}{n}\right)$$

$$= -\sum_{k=1}^{\frac{n}{2}-1} \cos\left(\frac{kj\pi}{n}\right)$$

since for an odd j, $\cos(j\pi - \alpha) = -\cos(\alpha)$. Relation (3.22) shows that the lemma is true in this case. Next assume that n is odd and write

$$\sum_{m=1}^{n-1} \cos\left(\frac{mj\pi}{n}\right) = \sum_{m=1}^{\frac{n-1}{2}} \cos\left(\frac{mj\pi}{n}\right) + \sum_{m=\frac{n+1}{2}}^{n-1} \cos\left(\frac{mj\pi}{n}\right). \quad (3.23)$$

Again, the same change of index $(k = n - m)$ in the second sum shows that it is equal to the opposite of the first one and the lemma is proved. □

3.5.4 *Proof of Theorem 4.1*

We are now ready to prove Theorem 4.1.

Consider two columns

$$q_i = \begin{bmatrix} \sqrt{\frac{1}{n}} \\ \sqrt{\frac{2}{n}}\cos\left(\frac{(2i+1)\pi}{2n}\right) \\ \sqrt{\frac{2}{n}}\cos\left(\frac{2(2i+1)\pi}{2n}\right) \\ \vdots \\ \sqrt{\frac{2}{n}}\cos\left(\frac{(n-1)(2i+1)\pi}{2n}\right) \end{bmatrix}, \quad q_j = \begin{bmatrix} \sqrt{\frac{1}{n}} \\ \sqrt{\frac{2}{n}}\cos\left(\frac{(2j+1)\pi}{2n}\right) \\ \sqrt{\frac{2}{n}}\cos\left(\frac{2(2j+1)\pi}{2n}\right) \\ \vdots \\ \sqrt{\frac{2}{n}}\cos\left(\frac{(n-1)(2j+1)\pi}{2n}\right) \end{bmatrix}$$

$(0 \le i, j \le n - 1)$ of the DCT matrix D displayed in Figure 3.1 on page 71. Without loss of generality, we can assume $i \ge j$. We compute the dot product of q_i and q_j:

$$q_i \cdot q_j = \sqrt{\frac{1}{n}}\sqrt{\frac{1}{n}} + \sum_{k=1}^{n-1} \sqrt{\frac{2}{n}} \cos\left(\frac{k(2i+1)\pi}{2n}\right) \sqrt{\frac{2}{n}} \cos\left(\frac{k(2j+1)\pi}{2n}\right)$$

$$= \frac{1}{n} + \frac{2}{n}\sum_{k=1}^{n-1} \cos\left(\frac{k(2i+1)\pi}{2n}\right) \cos\left(\frac{k(2j+1)\pi}{2n}\right)$$

$$= \frac{1}{n} + \frac{2}{n}\sum_{k=1}^{n-1} \frac{1}{2}\left[\cos\left(\frac{k(2i+1)\pi}{2n} + \frac{k(2j+1)\pi}{2n}\right)\right.$$

$$\left. + \cos\left(\frac{k(2i+1)\pi}{2n} - \frac{k(2j+1)\pi}{2n}\right)\right] \text{ (by (3.19))}$$

$$= \frac{1}{n} + \frac{1}{n}\sum_{k=1}^{n-1} \cos\left(\frac{k(i+j+1)\pi}{n}\right) + \frac{1}{n}\sum_{k=1}^{n-1} \cos\left(\frac{k(i-j)\pi}{n}\right)$$

with $1 \leq i+j+1 \leq 2n-1$ and $0 \leq i-j \leq n-1$. In order for the DCT matrix D to be orthogonal, two things have to be verified. First, if $i = j$, then $q_i \cdot q_j$ must be 1. Second, if $i \neq j$, then $q_i \cdot q_j$ must be 0. The case $i = j$ is easier to treat and that is what we start with. Note that in this case, $\cos\left(\frac{k\pi(i-j)}{n}\right) = 1$ for any k which means that $\sum_{k=1}^{n-1} \cos\left(\frac{k\pi(i-j)}{n}\right) = n-1$. On the other hand, $\sum_{k=1}^{n-1} \cos\left(\frac{k\pi(i+j+1)}{n}\right) = \sum_{k=1}^{n-1} \cos\left(\frac{k\pi(2i+1)}{n}\right) = 0$ by Lemma 3.3. We conclude then that $q_i \cdot q_j = \frac{1}{n} + \frac{1}{n}(n-1) = 1$ for $i = j$.

Assume next that $i \neq j$ so q_i, q_j are distinct columns of D. Note that if $i - j = 2\alpha$ for some $\alpha \in \mathbb{Z}$ (that is if $i-j$ is even), then $i+j+1 = 2j+2\alpha+1$ is odd. Similarly, if $i-j$ is odd then $i+j+1$ is even. The proof that the dot product of q_i and q_j is zero in this case is done by considering the following two cases.

- $i - j$ is even. In this case

$$\frac{1}{n}\sum_{k=1}^{n-1} \cos\left(\frac{k\pi(i-j)}{n}\right) = \overbrace{\frac{1}{n}\sum_{k=0}^{n-1} \cos\left(\frac{k\pi(i-j)}{n}\right)}^{=0 \text{ by Lemma 3.2}} - \frac{1}{n} = -\frac{1}{n}.$$

On the other hand, Lemma 3.3 shows that $\sum_{k=1}^{n-1} \cos\left(\frac{k\pi(i+j+1)}{n}\right) = 0$ since $i + j + 1$ is odd in this case. Therefore, $q_i \cdot q_j = \frac{1}{n} - \frac{1}{n} = 0$.

- $i - j$ is odd. Interchanging the roles of $i-j$ and $i+j+1$ in the previous case shows that $q_i \cdot q_j = 0$ again in this case. By Theorem 3.1, the DCT matrix D is indeed orthogonal.

3.6 References

Brown, S. and Vranesic., Z. (2005). *Digital Logic with VHDL Design,* Second Edition. (McGraw-Hill).

Predko, M. (2005) *Digital electronic demystified.* (McGraw-Hill).

Farhat, H.A. (2004) *Digital design and computer organization.* (CRC Press).

Chapter 4

Global Positioning System (GPS)

4.1 Introduction

Recently, after a family trip, a friend of mine decided to go back to use his old paper map in his travels and to put his car GPS to rest forever. This came after a series of deceptions by this little device with the annoying automated voice, the $4''$ screen and the frequently interrupted signal (his words not mine). The latest of these deceptions was a trip from Ottawa to Niagara Falls which took a turn in the US. Admittedly, such a turn is normal especially if the GPS is programmed to take the shortest distance, except that my friend's family did not have passports on them that day.

If you have used a GPS before, you must have experienced some set-backs here and there. But let us face it, the times when the trip goes smoothly without any wrong turns or lost signal, we cannot help but admire the magic and ingenuity that transforms a little device into a holding hand that takes you from point A to point B, sometimes thousands of kilometers apart. It is almost "spooky" to think that someone is watching you every step of the way from somewhere "out of this Earth".

In this chapter, you will learn that there is nothing magical about the GPS. It is the result of collective efforts of scientists and engineers with Mathematics as the main link. After reading this chapter, use your time on the road in your next trip to try to reveal to your co-travelers (with as little mathematics as possible) the secret behind this technology. It works every time I want to put my kids to sleep on a long trip.

4.1.1 *Before you go further*

Although the chapter is intended to be self-contained as much as possible, it is the heaviest in mathematical content compared to other chapters. The structure of the GPS signals involves the use of abstract mathematical concepts, chiefly from abstract algebra like group theory, finite fields, polynomial rings and primitive elements. Readers who want to have a full grasp on the mathematical proofs are expected to read through more than one time.

4.2 Latitude, longitude and altitude

If you press the "where am I" or "My location" buttons built in your GPS receiver, your location will be displayed with expressions like 40°N, 30°W and 1040 m, which are clearly not the "classical" cartesian coordinates. This is because your GPS uses a more efficient coordinate system in which the position or location of any point on or near the earth's surface is determined by three parameters known as the *latitude*, the *longitude* and the *altitude*. You may have probably seen these terms before in a geography class, but let us review them anyway. First, consider a cartesian coordinate system $Oxyz$ of three orthogonal axes centered at the center O of the earth.

Consider a point $Q(x, y, z)$ in the above coordinate system and let P be the projection of Q on the earth surface. That is, P is the intersection point of the vector $\overrightarrow{OQ}$ with the earth surface. The points Q and P share the same latitude and longitude that we define in what follows.

- the **latitude** of P (same as the latitude of Q) is the measurement of the angle β of the location of P north or south of the equator. It represents the angle formed between the position vector $\overrightarrow{OP}$ and the plane of the equator. Note that $-90° \leq \beta \leq 90°$ with the point of latitude $-90°$ being the South Pole that we mark as 90°S and the point of latitude 90° being the North Pole that we mark as 90°N. Points of latitude 0° are points on the Equator. Lines of latitude (known as **parallels**) are (imaginary) circles on the planet surface each parallel to the equator plane. Points on the same parallel share the same latitude.
- A **meridian line** is an imaginary north-south circle on the earth's surface connecting the north and south poles. The meridian line passing through the town of Greenwich, England is agreed on internationally to

be a reference line known as the **prime meridian** (also known as the Greenwich line). The **longitude** of the point P (same as the longitude of Q) is the measurement of the angle ϕ of the location of P east or west of the prime meridian. Note that $-180° \leq \phi \leq 180°$ with points of negative longitude are located to the west of the prime meridian and points with positive longitude are to its east. Thus a longitude of $-100°$ is written as $100°W$ and a longitude of $55°$ is written as $55°E$. Points on the same meridian share the same longitude. Note that meridian lines are orthogonal to parallel lines.

- The **altitude** h of Q is its distance from the sea level. If R is the radius of the earth ($R \cong 6366\,\text{km}$), then the distance between the point Q and the center of the earth is $R + h$.

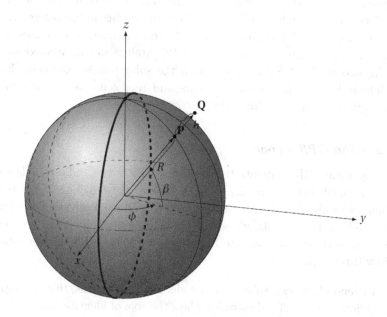

The position of any point near the surface of the planet is uniquely determined by its latitude, it longitude and its altitude. For example, a point described as ($40°N$, $30°W$, $1850\,\text{m}$) is a point located $40°$ north of the Equator and $30°$ west of the Greenwich meridian and at a distance of 1.85 km from the sea level (or $6366 + 1.85 = 6367.85$ km from the center of the earth).

4.3   About the GPS system

4.3.1   *The GPS constellation*

The GPS system is a constellation of man-made satellites placed in six equally-spaced orbital planes surrounding the Earth. Each orbital plane is inclined 55 degrees relative to the plane of the equator. There are at least 24 operational satellites at any given time with a number of backup satellites in case of failures. The 24 operational satellites are arranged in four satellites per orbit. Each GPS satellite orbits the earth twice a day (once every 12 hours) at an altitude of approximately 20,200 km (12,550 miles). The number of operating satellites in each orbit, altitude and inclination of their orbital planes as well as their speed and distances apart are carefully chosen to ensure that a GPS receiver will always have in its range at least four satellites no matter where it is located near the surface of the planet. As to why we need four satellites in the range of a receiver, the answer comes a bit later. In case you are interested, each GPS satellite weighs approximately 908 kg and is about 5.2 m across with the solar panels extended. Each satellite is built to last about 10 years and replacements are constantly being built and launched into orbits.

4.3.2   *The GPS signal*

Each operating GPS satellite transmits constantly two radio signals with frequencies labeled L_1 and L_2. The frequency L_1 (1575.42 MHz) is for civilian use while L_2 can only be depicted by military receivers. The signal can travel through clouds, glass and plastic but it is reflected by objects like water surfaces and concrete buildings. A typical GPS signal contains mainly three segments:

- A Pseudo Random Noise (PRN) code. In simple terms, this is a digital sequence of on/off pulses which plays the role of identification code for the satellite transmitting the signal. Each GPS satellite generates its own and unique PRN and the code is designed with enough complexity to make it virtually impossible for a ground receiver to confuse the signal with another from outside the constellation. The uniqueness of the PRN allows all operating satellites to use the same frequency.

- An ephemeris data. This part of the signal contains detailed information about the orbit of the satellite and where it should be at any given time in addition to the current date and time according to the (atomic) clock on board of the satellite. The information contained in this part of the signal is vital for the operation of the receiver.

- An almanac data. This part provides the GPS receiver with information about *all* the operating GPS satellites in the constellation. Each satellite emits almanac data about its own orbit as well as other satellites. With this information, the receiver can determine which satellites are likely to be in its range and does not waste time looking for the ones that are not.

4.4 Pinpointing your location

Your GPS receiver uses a simple mathematical concept called *Trilateration* to locate its position at any given time. We start by explaining this principle in the case of a "two-dimensional" map.

4.4.1 *Where am I on the map?*

Imagine you are lost on campus, you are holding a campus map in your hand but you do not feel it offers much help. You ask someone: "Where am I?" and the person answers, "You are 500 m away from the university center" and he walks away. You locate the university center, labeled as UC on the campus map, but that does not help much since you could be anywhere on the circle C_1 centered at UC and of radius 500 m. You draw C_1 using the scaling of the campus map.

Next, you ask the same question to another person passing by, and he answers: "You are 375 m away from the Math Department" and walks away. You locate the Math Department on the map, labeled as MD, and you draw on your map the circle C_2 centered at MD and of radius 375 m. This new information narrows your location to two possible points A and

B, namely the intersection points of circles C_1 and C_2.

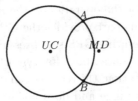

To know which of the two points A and B is your exact location, it suffices to draw a third circle that would intersect the other two at one of these two points. You locate another building relatively close to UC and MD, say the Faculty of Engineering, labeled as FE on the map. You ask a third person passing by: "How far am I from the Faculty of Engineering?" and he answers, "About 200 m". You then draw the circle C_3 on the map centered at FE and of radius 200 m.

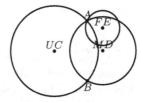

The point where the three circles meet determines your (relatively) exact location.

Of course, in order for this to work, you must be lucky enough to have people passing by giving you (relatively) precise distances from various locations and to be able to somehow work the scale of the map to draw accurate circles. Equally important is the kind of question you should ask the third person in order to ensure that the third circle will somehow meet the other two at exactly one point. Roughly speaking, a GPS receiver works the same way except that the circles are replaced by spheres in three dimensions, and the friendly people you ask to pinpoint your position on the campus map are replaced with satellites located thousands of kilometers above the surface of the earth.

4.4.2 *Measuring the distance to a satellite*

Now for the story of locating your position on the surface of the planet.

Signals transmitted by GPS satellites travel at the speed of light (at least in a vacuum) and reach the GPS receiver at slightly different times as some satellites are further away from the receiver than others. These signals are repeated continuously and any GPS receiver has them stored in its internal memory along with what the value of the sequence should be at any given time. As the receiver repeats the satellites sequences internally, the captured sequence and the one generated by the receiver must be in synchronize mode in theory, but they are not because of the time taken by the signal to reach the receiver. Once the receiver captures a signal, it immediately recognizes which satellite it is coming from (using the PRN segment of the signal) and its compares it to its own replica. By comparing how much the satellite signal is lagging, the travel time dt of the signal to reach the receiver is calculated.

Does this seem to be a bit too technical? In the next paragraph we try to explain the idea of the "time lag" using a simple example.

Let us assume that a GPS satellite signal is just a "song" broadcasted by the satellite (admittedly not a pretty one). Imagine that at 6:00 am, a GPS satellite begins to broadcast the song "*I see trees of green, red roses too, I see them bloom for me and you...*" in the form of a radio wave to Earth. At exactly the same time, a GPS receiver starts playing the same song. After traveling thousands of kilometers in space, the radio wave arrives at the receiver but with a certain delay in the words. If your standing by the receiver, you will hear two interfering versions of the song at the same time. At the time of signal reception, the receiver version is playing "*...them bloom for...*" but the satellite version is playing (for instance) the first "*I see...*". The receiver player would then immediately "rewind" its version a bit until it synchronizes perfectly with the received version. The amount of time equivalent to this "shift back" in the receiver player is precisely the travel time of the satellite's version.

Once the time delay dt is computed, the receiver internal computer multiplies it with the speed of light (in a vacuum), $c = 299,792,458\,\text{m/sec}$ to calculate the distance separating the satellite from the GPS receiver.

4.4.3 *Where am I on the surface of the planet?*

Now that we have a bit more understanding of how the GPS receiver estimates its distance to the satellites in its view, it is time to see how these estimates are put in use to pinpoint the position of the receiver.

We start by choosing a system of three orthogonal axes centered at the point O, the center of the earth. The z-axis is the vertical one passing through the two poles and oriented from South to North. The xz plane is the Greenwich meridian plane. The x-axis lies in the equatorial plane and the direction of positive values of x goes through the Greenwich point (point of longitude zero). Similarly, the y-axis lies in the equatorial plane and the direction of positive values of y goes through the point of longitude 90° East.

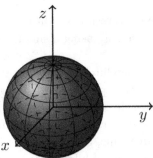

All GPS receivers are built with multiple channels allowing them to receive and treat signals from at least four different satellites simultaneously. Once it captures the signals of three satellites S_1, S_2 and S_3 in its range, the receiver calculates the time delays t_1, t_2 and t_3 (respectively, in seconds) taken by signals of the three satellites to reach it. The distances between the receivers and the three satellites are computed as explained in the previous section: $d_1 = ct_1, d_2 = ct_2$, and $d_3 = ct_3$, respectively. The fact that the receiver is at a distance d_1 from satellite S_1 means that it could be anywhere on the (imaginary) sphere Σ_1 centered at S_1 and of radius d_1. Using the ephemeris data part of the satellite signal, the receiver knows the position (a_1, b_1, c_1) of the satellite S_1 in the above system of axes, so the sphere Σ_1 has equation:

$$(x - a_1)^2 + (y - b_1)^2 + (z - c_1)^2 = d_1^2 = c^2 t_1^2. \tag{4.1}$$

The distance $d_2 = ct_2$ from the second satellite is computed and the receiver is also somewhere on the sphere Σ_2 centered at the satellite S_2, positioned at the point (a_2, b_2, c_2), with radius d_2:

$$(x - a_2)^2 + (y - b_2)^2 + (z - c_2)^2 = d_2^2 = c^2 t_2^2. \tag{4.2}$$

This narrows the position of the receiver to the intersection of two spheres, namely to a circle Γ. Still not enough to determine the exact position. Finally, the distance $d_3 = ct_3$ from the third satellite S_3, positioned at the point (a_3, b_3, c_3), shows that the receiver is also on the sphere Σ_3:

$$(x - a_3)^2 + (y - b_3)^2 + (z - c_3)^2 = d_3^2 = c^2 t_3^2. \tag{4.3}$$

The inclination of orbits of GPS satellites are designed so that the surface of the third sphere would intersect Γ in two points that the receiver can accurately compute their coordinates. One of these two points will be unreasonably far from the surface of the earth and therefore one possible position is left.

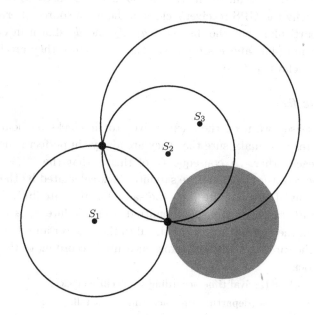

4.4.4 *Is it really that simple?*

In theory, once a GPS receiver captures the signals of three different satellites in its view, it should be able to locate its exact position (as the intersection of three imaginary spheres). But in reality, things are bit more complicated than that.

The calculation of the time taken by the satellite signal to reach the receiver (as explained above) assumes that clocks in the receiver and on board of the satellite are in perfect synchronization. So 6:00 am on board of the satellite means 6:00 am on the receiver clock. Unfortunately, that is not the case. The satellites are equipped with atomic clocks, very sophisticated, and extremely accurate clocks, but very expensive. The clocks inside the receivers, on the other hand, are the usual everyday digital clock. The difference between the types of clocks creates a certain error in calculating the real time delay of the GPS signal. You may wonder, why the big fuss about a time estimate that could differ only in a fraction of a second? Remember, we are talking about waves traveling at the speed of light which makes the estimated distances from the satellite to the GPS receiver extremely sensitive to gaps between the satellite and receiver clocks. To give you an idea about the degree of sensitivity, an error of 0.000001 second (one microsecond) would result in an error of 300 metres in distance estimation. No wonder why the GPS receiver's clock is the main source of error. This means in particular that the distances d_1, d_2 and d_3 shown in equations (4.1), (4.2) and (4.3) above are not very accurate since they are based on "fake" time delays t_1, t_2 and t_3.

4.4.5   *The fix*

The main reason we need these expensive atomic clocks on board of the GPS satellites is to make sure that they are always in perfect synchronization with each other. A consequence of this is that the "time error" ξ between the receiver clock and the satellite clock calculated by the receiver is independent of the satellite (i.e., the same for any satellite). The **real** travel time dt_i taken by the signal emitted from satellite S_i is the difference between the arrival time of the signal to the receiver and its departure time from the satellite with both times measured according to the satellite (atomic) clock:

$$
\begin{aligned}
dt_i = &\ (\text{arrival time according to satellite clock}) \\
&- (\text{departure time according to satellite clock}) \\
= &\ (\text{arrival time according to receiver clock}) \\
&- (\text{departure time according to satellite clock}) \\
&- [(\text{arrival time according to receiver clock}) \\
&- (\text{arrival time according to satellite clock})] \\
= &\ t_i - \xi.
\end{aligned}
$$

The true travel time of the signal from satellite S_i is then equal to $t_i - \xi$ (with t_i as above) rather than simply t_i. Equations (4.1), (4.2) and (4.3) above can now be written as:

$$(H) \begin{cases} (x - a_1)^2 + (y - b_1)^2 + (z - c_1)^2 = d_1^2 = c^2(t_1 - \xi)^2 \\ (x - a_2)^2 + (y - b_2)^2 + (z - c_2)^2 = d_2^2 = c^2(t_2 - \xi)^2 \\ (x - a_3)^2 + (y - b_3)^2 + (z - c_3)^2 = d_3^2 = c^2(t_3 - \xi)^2 \end{cases}.$$

This is a system of three equations in four unknowns: the three coordinates of the receiver position (x, y and z) and the clocks offset time ξ. To solve the system for x, y and z, one possibility is to eliminate the need for the fourth variable ξ all together by equipping the receivers with atomic clocks so they perfectly synchronize with the satellites clocks. That would reduce ξ to zero in the system (H) giving a system of three equations in three unknowns that the receiver computer can solve to figure out its position. Of course, that would mean paying tens of thousands of dollars for the receiver. Not a smart way to make this technology available to the general public. So how come almost everyone you know has a very affordable GPS receiver that is very accurate at the same time? The designers of the GPS came up with an answer that is mathematically brilliant and yet simple. As it turns out, a simple digital clock in your GPS receiver will do just fine and all what it takes is one more measurement from a fourth satellite and voilà, you have the equivalent of an atomic clock right in the palm of your hand.

As explained earlier, the GPS satellites are placed in orbits so that there are always at least four satellites in view of a GPS receiver anywhere near the surface of the planet. The receiver captures the signal of a fourth satellite S_4 and adds one more equation to the above system (H). Now we have the following system of four equations in four unknowns to deal with:

$$(S) \begin{cases} (x - a_1)^2 + (y - b_1)^2 + (z - c_1)^2 = d_1^2 = c^2(t_1 - \xi)^2 \\ (x - a_2)^2 + (y - b_2)^2 + (z - c_2)^2 = d_2^2 = c^2(t_2 - \xi)^2 \\ (x - a_3)^2 + (y - b_3)^2 + (z - c_3)^2 = d_3^2 = c^2(t_3 - \xi)^2 \\ (x - a_4)^2 + (y - b_4)^2 + (z - c_4)^2 = d_4^2 = c^2(t_4 - \xi)^2 \end{cases}.$$

4.4.6 Finding the coordinates of the receiver

Note first that the system (S) is not a linear system and solving it would require more than the techniques seen in a basic linear algebra course. But with a little effort, it could be brought to a "quasi-linear" form. We start by subtracting the fourth equation in (S) from each of the first three equations. For instance, subtracting the fourth equation from the first gives:

$$(x - a_1)^2 + (y - b_1)^2 + (z - c_1)^2 - \left[(x - a_4)^2 + (y - b_4)^2 + (z - c_4)^2\right]$$
$$= c^2(t_1 - \xi)^2 - c^2(t_4 - \xi)^2.$$

This results in the following equation:

$$2(a_4 - a_1)x + 2(b_4 - b_1)y + 2(c_4 - c_1)z = 2c^2(t_4 - t_1)\xi + (a_4^2 + b_4^2 + c_4^2)$$
$$- (a_1^2 + b_1^2 + c_1^2) - c^2(t_4^2 - t_1^2).$$

The expression $(a_4^2 + b_4^2 + c_4^2) - (a_1^2 + b_1^2 + c_1^2) - c^2(t_4^2 - t_1^2)$ in the above equation is *independent* of the variables x, y, z and ξ of the system. To simplify the notations a little bit, we call it A_1:

$$A_1 = (a_4^2 + b_4^2 + c_4^2) - (a_1^2 + b_1^2 + c_1^2) - c^2(t_4^2 - t_1^2).$$

This way, the last equation can now be written as:

$$2(a_4 - a_1)x + 2(b_4 - b_1)y + 2(c_4 - c_1)z = 2c^2(t_4 - t_1)\xi + A_1. \quad (4.4)$$

Repeating the same thing for the second and third equations in (S), we obtain a new system (S') equivalent to (S) (in the sense that both systems have the same set of solutions):

$$(S') \begin{cases} 2(a_4 - a_1)x + 2(b_4 - b_1)y + 2(c_4 - c_1)z = 2c^2(t_4 - t_1)\xi + A_1 \\ 2(a_4 - a_2)x + 2(b_4 - b_2)y + 2(c_4 - c_2)z = 2c^2(t_4 - t_2)\xi + A_2 \\ 2(a_4 - a_3)x + 2(b_4 - b_3)y + 2(c_4 - c_3)z = 2c^2(t_4 - t_3)\xi + A_3 \\ (x - a_4)^2 + (y - b_4)^2 + (z - c_4)^2 = d_4^2 = c^2(t_4 - \xi)^2 \end{cases}.$$

One way to solve (S') is to treat ξ as a constant in each of the first three equations. This will allow us to express each of the variables x, y and z in terms of ξ and then use the fourth equation to find ξ (hence x, y and z). This approach enables us to use the techniques of linear algebra to solve systems of linear equations since the first three equations in (S') form indeed a system of three linear equations in three variables (x, y and z).

There are many ways to solve for x, y and z in term of ξ in the first three equations, but Cramer's rule is probably the easiest to implement in the receiver's computer:

$$x = \frac{D_1}{D}, \quad y = \frac{D_2}{D}, \quad z = \frac{D_3}{D},$$

.

where D is the determinant of the matrix:

$$L := \begin{bmatrix} 2(a_4 - a_1) & 2(b_4 - b_1) & 2(c_4 - c_1) \\ 2(a_4 - a_2) & 2(b_4 - b_2) & 2(c_4 - c_2) \\ 2(a_4 - a_3) & 2(b_4 - b_3) & 2(c_4 - c_3) \end{bmatrix}$$

and D_1, D_2, D_3 are respectively the determinants of the matrices

$$L_1 = \begin{bmatrix} 2c^2(t_4 - t_1)\xi + A_1 & 2(b_4 - b_1) & 2(c_4 - c_1) \\ 2c^2(t_4 - t_2)\xi + A_2 & 2(b_4 - b_2) & 2(c_4 - c_2) \\ 2c^2(t_4 - t_3)\xi + A_3 & 2(b_4 - b_3) & 2(c_4 - c_3) \end{bmatrix},$$

$$L_2 = \begin{bmatrix} 2(a_4 - a_1) & 2c^2(t_4 - t_1)\xi + A_1 & 2(c_4 - c_1) \\ 2(a_4 - a_2) & 2c^2(t_4 - t_2)\xi + A_2 & 2(c_4 - c_2) \\ 2(a_4 - a_3) & 2c^2(t_4 - t_3)\xi + A_3 & 2(c_4 - c_3) \end{bmatrix},$$

$$L_3 = \begin{bmatrix} 2(a_4 - a_1) & 2(b_4 - b_1) & 2c^2(t_4 - t_1)\xi + A_1 \\ 2(a_4 - a_2) & 2(b_4 - b_2) & 2c^2(t_4 - t_2)\xi + A_2 \\ 2(a_4 - a_3) & 2(b_4 - b_3) & 2c^2(t_4 - t_3)\xi + A_3 \end{bmatrix}.$$

Clearly, we would be in trouble if $D = 0$. But can that really happen? Using the properties of determinants, we can write

$$D = 8 \begin{vmatrix} a_4 - a_1 & b_4 - b_1 & c_4 - c_1 \\ a_4 - a_2 & b_4 - b_2 & c_4 - c_2 \\ a_4 - a_3 & b_4 - b_3 & c_4 - c_3 \end{vmatrix} \tag{4.5}$$

(the 8 in front is obtained by factoring 2 from each of the three rows of D) where a_i, b_i, c_i are the coordinates of the satellite S_i in the above system of axes. So the rows in the determinant D are the components of the vectors $\overrightarrow{S_1 S_4}$, $\overrightarrow{S_2 S_4}$ and $\overrightarrow{S_3 S_4}$ respectively (the S_i's being the satellites). If $D = 0$, then a known result from linear algebra asserts that the three vectors are coplanar (belong to the same orbital plane) and consequently, the four satellites S_1, S_2, S_3 and S_4 are also coplanar. Engineers were of course fully aware of this problem and the way they chose to place the 24 satellites in their orbits was carefully chosen so that it makes it impossible for a GPS receiver to capture the signals of four coplanar satellites at any moment and anywhere close to the surface of the Earth. Your linear algebra course does not look so abstract now, does it?

Replacing x, y and z by $\frac{D_1}{D}$, $\frac{D_2}{D}$ and $\frac{D_3}{D}$ respectively in the fourth equation of (S') yields the following quadratic equation

$$\left(\frac{D_1}{D} - a_4 \right)^2 + \left(\frac{D_2}{D} - b_4 \right)^2 + \left(\frac{D_3}{D} - c_4 \right)^2 = c^2(t_4 - \xi)^2$$

which can be written as

$$c^2\xi^2 - 2c^2t_4\xi + \kappa = 0 \tag{4.6}$$

where $\kappa = c^2t_4^2 - \left(\frac{D_1}{D} - a_4\right)^2 - \left(\frac{D_2}{D} - b_4\right)^2 - \left(\frac{D_3}{D} - c_4\right)^2$. Once again, the way the satellites are put in their orbits guarantees that equation (4.6) would have two solutions ξ_1 and ξ_2. This gives two possible positions (one for each of the two values found for ξ) with one of them corresponding to a point very far from the surface of the planet that the receiver eliminates as a possibility.

4.4.7   Conversion from cartesian to (latitude, longitude, altitude) coordinates

At this point, the receiver knows its position in cartesian form as being the point $Q(x, y, z)$ in the above coordinates system $Oxyz$. But the display on the screen shows the position in the (latitude, longitude, altitude) coordinates. The conversion is done as in the following steps.

- First, note that the distance separating the receiver from the earth center is $d = \sqrt{x^2 + y^2 + z^2}$.
- The altitude of the receiver is computed as $h = d - R$ where R is the radius of the earth.
- For the point P, the projection of Q on the surface of the earth, the cartesian coordinates are given by $\frac{R}{d}x$, $\frac{R}{d}y$ and $\frac{R}{d}z$. The relations between these coordinates and the latitude β and the longitude ϕ of the point P (which are the same for the point Q) are given by:

$$\begin{cases} \frac{R}{d}x = R\cos\beta\cos\phi \\ \frac{R}{d}y = R\sin\phi\cos\beta \ . \\ \frac{R}{d}z = R\sin\beta \end{cases}$$

These can be simplified to the following equations:

$$(L)\begin{cases} x = d\cos\beta\cos\phi \\ y = d\sin\phi\cos\beta \ . \\ z = d\sin\beta \end{cases}$$

The last equation gives that $\sin\beta = \frac{z}{d}$ and since $-90° \le \beta \le 90°$, there is a unique value of β satisfying $\sin\beta = \frac{z}{d}$, namely $\beta = \arcsin\left(\frac{z}{d}\right)$ (or maybe you have seen this as $\sin^{-1}\left(\frac{z}{d}\right)$ in your first calculus course).

- Knowing the value of β, $\cos\beta$ can be computed and the system (L) is now reduced to the following two equations:

$$\begin{cases} \cos\phi = \frac{x}{d\cos\beta} \\ \sin\phi = \frac{y}{d\cos\beta} \end{cases}$$

(with $\cos\beta$ known). Since $-180° \le \phi \le 180°$, these two equations determine uniquely the value of the longitude ϕ.

- Thus the position $Q(x,y,z)$ of the receiver can now be displayed in terms of the latitude, the longitude and the altitude of the position point Q.

4.5 The mathematics of the GPS signal

Obviously, the satellites are not emitting their signals using the words of the song "*I see trees of green red roses too...*". So what is the nature of these signals and how are they engineered to be easily identified by a ground receiver? More importantly, how can we make sure the signal is sufficiently "random" to suit the intended use?

Locating the receiver position may have appeared somehow complicated to you, but the truth is this was the "soft" side of the mathematics used in this project. Careful encryption of codes in the signal emitted by the satellite is key to ensure accuracy and reliability of information provided to your receiver. This side of the GPS project requires somewhat heavier mathematics.

4.5.1 *Terminology*

We start by going over some of the terminology needed for the rest of this section. As seen in previous chapters, a **binary sequence** is a sequence consisting of only two symbols, usually denoted by of 0 and 1 (or On/Off pulses), that we call **bits**. A binary sequence is called of **length** r if it is a finite sequence consisting of r bits. A sequence a_0, a_1, a_2, ... is called **periodic** if there exists a positive integer p, called a **period** of the sequence, such that $a_{n+p} = a_n$ for all n. In other words, the periodic sequence repeats itself every cycle of p terms. Note that if p is a period, then kp is also a period for any positive integer k. The smallest possible value for p is called the **minimal period** (in some books, it is simply called the period) of the sequence. For example, the sequence

001011000101100010110001011000101100010110001011000101100010110

is a binary sequence of length 63 and it is periodic of (minimal) period 7 with the block 0010110 of 7 digits repeated nine times. Note that a binary sequence of length r can be viewed as a vector $(a_0, a_1, \ldots, a_{r-1})$ having r components with each component a_i is an element of the set $\mathbb{F}_2 := \{0, 1\}$. This means in particular that there are exactly 2^r different binary sequences of length r. For example, there are $2^3 = 8$ binary sequences of length 3, namely 111, 110, 101, 100, 011, 010, 001 and 000.

4.5.2   *Linear Feedback Shift Registers*

In their raw form, the pseudo random noise codes emitted by GPS satellites are exactly that: noise-like signals pretty much like the noise you hear when your radio cannot tune in to a station. The GPS receiver however is programmed to look beyond the noise. It treats the codes as deterministic binary sequences. The word "deterministic" in this context refers to the fact that for the receiver, these signals are not random but rather completely determined by a fixed set of coefficients and a relatively small number of initial values, called the PRNG's state (PRNG is an acronym for "Pseudo-Random Number Generator", the algorithm used to produce such a deterministic binary sequence). There are many pseudo-random number generators out there used for various applications but the one used in GPS signal is called **Linear Feedback Shift Register** or LFSR for short. In simple terms, a LFSR of *degree* m (or of m stages) can be described as a digital circuit containing a series of m one-bit storage (or memory) cells. Each cell is connected to a constant coefficient $c_i \in \{0, 1\}$. The vector $(c_0, c_1, \ldots, c_{m-1})$ is different from one satellite to another. Figure 4.1 below shows a block diagram of a typical LFSR. The rhythm of the register is controlled by a counter or a clock. The system is capable of generating a sequence a_n of binary bits that will have the "appearance" of being very random using the following steps.

(1) Start by choosing an initial "window" $w_0 = (a_0, a_1, \ldots, a_{m-1})$ (in the figure for the LFSR block diagram, values are read from right to left). These initial values are not all zeros at the same time ($w_0 \neq \mathbf{0}$) and the vector w_0 is different for different satellites.

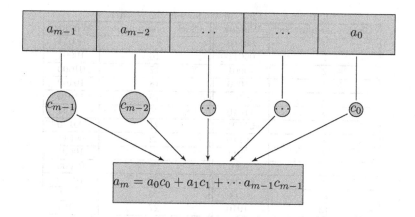

Fig. 4.1 Linear feedback shift register block diagram.

(2) At the first "clock pulse", the content of each cell is shifted to the right by one box "pushing out" the value a_0. The content of the first (leftmost) box is then calculated as follows: first compute the value of the expression $\sum_{k=0}^{m-1} a_k c_k = a_0 c_0 + a_1 c_1 + \cdots + a_{m-1} c_{m-1}$. If the result is even, the value $a_m = 0$ is inserted in the leftmost box. If the result is odd, the value $a_m = 1$ is inserted in the leftmost box. If you are familiar with modular arithmetic (see Section 4.5.3 below), this amounts to calculating the sum $\sum_{k=0}^{m-1} a_k c_k$ "modulo" 2. We now have the second "window" $w_1 = (a_1, a_2, \ldots, a_m)$ (again read from right to left in the block diagram) and the first $m + 1$ terms of the sequence are $a_m, a_{m-1}, \ldots, a_1, a_0$.

(3) This process is repeated. For example, at the second "clock pulse", the register shifts again the content of each cell to the right by one box pushing out the value a_1 this time. The content of the leftmost box is calculated as $c_0 a_1 + c_1 a_2 + \cdots + c_{m-2} a_{m-1} + c_{m-1} a_m$ (modulo 2) and this is precisely the next bit, a_{m+1}, in the sequence. We now have the following terms $a_{m+1}, a_m, a_{m-1}, \ldots, a_1, a_0$ of the sequence.

(4) The procedure is iterated, creating an infinite binary sequence $\ldots, a_k, a_{k-1}, \ldots, a_2, a_1, a_0$.

Before we dig deeper in the mathematical properties of the sequence produced by a LSFR, let us look at a simple example.

Example 4.1. In this example, we consider a LFSR of degree 5 ($m = 5$). As coefficient vector, we take $c = (c_0, c_1, c_2, c_3, c_4) = (0, 1, 1, 1, 0)$ and as

Table 4.1   First 30 windows in the register.

Clock pulse number	Window	Clock pulse number	Window
0	00110	15	00011
1	00011	16	10001
2	10001	17	01000
3	01000	18	10100
4	10100	19	11010
5	11010	20	01101
6	01101	21	00110
7	00110	22	00011
8	00011	23	10001
9	10001	24	01000
10	01000	25	10100
11	10100	26	11010
12	11010	27	01101
13	01101	28	00110
14	00110	29	00011

initial state we take the vector $w_0 = (a_0, a_1, a_2, a_3, a_4) = (0, 1, 1, 0, 0)$ (or 00110 when the values in the window are read from left to right). At the first clock pulse, the register computes the sum $(0 \times 0) + (1 \times 1) + (1 \times 1) + (1 \times 0) + (0 \times 0) = 2$. Since the result is even, the content of the leftmost box is 0. The new window in the sequence is 00011 (again from left to right). At the second clock pulse, the register computes the sum $(0 \times 1) + (1 \times 1) + (1 \times 0) + (1 \times 0) + (0 \times 0) = 1$. Since the result is odd, the content of the leftmost box is 1 and the new window is 10001. Refer to Table 4.1 to find the first 30 windows in the register and the resulting sequence is $a_{28}a_{27} \cdots a_1 a_0 = 01000110100011010001101000110$.

Remark 4.1. Since there are exactly 2^m different binary sequences of length m, the sequence produced by a LFSR of degree m must be periodic of maximal period of 2^m. If you are not convinced, just look at the 30 windows produced by the LFSR in Example 4.1 above. Each window is a binary sequence of length 5, so there are $2^5 = 32$ (different) possible windows, including the window of zeros. In the worst case scenario, one needs 31 clock pulses before repeating a window and as soon as a window is repeated, the ones that follow will be already on the list in the same order. But note that the LSFR in Example 4.1 repeats the first window just after the seventh clock pulse. This justifies the notion of a "maximal period" of 2^m. Note also that no window of zeros appears in the table of Example 4.1 and this is no coincidence. If the coefficients $c_0, c_1, \ldots, c_{m-1}$ and the initial conditions $a_0, a_1, \ldots, a_{m-1}$ are "wisely" chosen (we will see how later), we can guarantee that no window of all zeros will ever occur and that the sequence produced by the register is periodical of maximum period possible.

All the mathematical machinery developed in the following sections is geared toward proving the following main result.

Theorem 4.1. For a LFSR of degree m, one can always choose the coefficients $c_0, c_1, \ldots, c_{m-1}$ and initial conditions $a_0, a_1, \ldots, a_{m-1}$ in such a way that the sequence produced by the register has a minimal period of maximal length $2^m - 1$.

4.5.3 *Some modular arithmetic*

The Division algorithm is one of those things we learn at early age in school but we usually do not pay much attention to its proper statement, let alone the mathematics behind it. It is a building block for almost everything we do in Arithmetic.

Theorem 4.2. (Division algorithm for integers) Given two integers a and b, with $b > 0$, there exist unique integers q and r, with $0 \leq r < b$, such that $a = bq + r$.

In the above theorem, q is called the **quotient**, r the **remainder**, b the **divisor** and a is called the **dividend** of the division.

Remark 4.2. When dividing an integer a by a non-zero integer b, one can always assume that $b > 0$ (if not, just divide $-a$ with $|b|$). Note that if $b < 0$, then $a = |b|q + r = b(-q) + r$ with $0 \leq r < |b|$ by the division algorithm.

For the rest of this section, we fix an integer $n \geq 2$.

Definition 4.1. We say that the two integers a and b are **congruent modulo** n and we write $a \equiv b \pmod{n}$, if a and b have the same remainder upon division by n.

If $a, b \in \mathbb{Z}$ have the same remainder upon division by n, then by the division algorithm we can write $a = nq_1 + r$ and $b = nq_2 + r$ for some q_1, q_2 and $r \in \mathbb{Z}$ with $0 \leq r < n$. So $a - b = n(q_1 - q_2)$ is divisible by n. Conversely, suppose that $a - b = \alpha n$ is divisible by n and write $a = nq_1 + r_1$ and $b = nq_2 + r_2$ for some q_1, q_2, r_1 and $r_2 \in \mathbb{Z}$ with $0 \le r_1 < n$ and $0 \le r_2 < n$. We can clearly assume that $r_2 \le r_1$ with no loss of generality (if not, just inverse the roles of a and b). So, $0 \le r_1 - r_2 < n$ and $a - b = n(q_1 - q_2) + (r_1 - r_2) = \alpha n$. By the uniqueness of the quotient and the remainder (Theorem 4.2), we conclude that $r_1 - r_2 = 0$. In other words, a and b have the same remainder upon division by n. This proves the following.

Theorem 4.3. For $a, b \in \mathbb{Z}$, $a \equiv b \pmod{n}$ if and only if $a - b$ is divisible by n.

As an example, $11 \equiv 21 \pmod 5$ since 11 and 21 have the same remainder (namely 1) upon division by 5 (or equivalently, their difference $21 - 11 = 10$ is divisible by 5).

There are n possible remainders upon division by n, namely $0, 1, \ldots, n - 1$. Given any integer a, the division algorithm allows us to write $a = nq + r$ for some $q, r \in \mathbb{Z}$ with $0 \le r \le n - 1$. Since $a - r = nq$ is divisible by n, we have that $a \equiv r \pmod{n}$. This means that any integer is congruent modulo n to one of the elements in the set $R = \{0, 1, \ldots, n - 1\}$. If $k \in R$, the set $\overline{k}$ of all integers having k as remainder in the division by n is known as an **equivalence class modulo** n:

$$\overline{k} := \{j \in \mathbb{Z}; \; j \equiv k \pmod{n}\}.$$

We then consider the collection $\mathbb{Z}_n$ of all equivalence classes modulo n:

$$\mathbb{Z}_n := \{\overline{k}; \; 0 \le k \le n - 1\}.$$

Example 4.2. $\mathbb{Z}_3 = \{\overline{0}, \overline{1}, \overline{2}\}$ where

$$\overline{0} = \{\ldots, -9, -6, -3, 0, 2, 6, 9, \ldots\},$$
$$\overline{1} = \{\ldots, -8, -5, -2, 1, 4, 7, 10, \ldots\},$$
$$\overline{2} = \{\ldots, -7, -4, -1, 2, 5, 8, 11, \ldots\}.$$

Remark 4.3. In the notation of the equivalence class $\overline{k}$ used above, the integer k is just one *representative* of that class. Any other element of the same class is also a representative. For instance, $\overline{1}$ can also be represented by $\overline{-5}$ or by $\overline{7}$ in $\mathbb{Z}_3$. To avoid confusion, the elements of $\mathbb{Z}_n$ are always represented in the (*standard*) form $\overline{k}$ for $0 \le k \le n - 1$. This way, we write $\overline{2}$ instead of $\overline{14}$ in $\mathbb{Z}_3$. It is also worth mentioning that $\overline{n} = \overline{0}$ since the remainder in the division of n by n is 0.

Our next task is to give the set $\mathbb{Z}_n$ a certain algebraic structure by defining an addition and a multiplication on the elements of the set that we call **addition and multiplication modulo** n. These operations are introduced naturally in the following way:

- **Addition modulo** n. If $\bar{a}$, $\bar{b} \in \mathbb{Z}_n$, define $\bar{a} + \bar{b}$ to be the equivalence class represented by the integer $a + b$. In other words, $\bar{a} + \bar{b} = \overline{a + b}$.
- **Multiplication modulo** n. If $\bar{a}$, $\bar{b} \in \mathbb{Z}_n$, define $\bar{a}\bar{b}$ to be the equivalence class represented by the integer ab: $\bar{a}\bar{b} = \overline{ab}$.

Since a class in $\mathbb{Z}_n$ has infinitely many representatives, one has to check that these two operations are independent of the choice of representatives. This is not hard to verify. The reader is certainly encouraged to try this as an exercise.

Example 4.3. The following are addition and multiplication tables of $\mathbb{Z}_3$:

+	$\bar{0}$	$\bar{1}$	$\bar{2}$
$\bar{0}$	$\bar{0}$	$\bar{1}$	$\bar{2}$
$\bar{1}$	$\bar{1}$	$\bar{2}$	$\bar{0}$
$\bar{2}$	$\bar{2}$	$\bar{0}$	$\bar{1}$

$\times$	$\bar{0}$	$\bar{1}$	$\bar{2}$
$\bar{0}$	$\bar{0}$	$\bar{0}$	$\bar{0}$
$\bar{1}$	$\bar{0}$	$\bar{1}$	$\bar{2}$
$\bar{2}$	$\bar{0}$	$\bar{2}$	$\bar{1}$

and of $\mathbb{Z}_4$:

+	$\bar{0}$	$\bar{1}$	$\bar{2}$	$\bar{3}$
$\bar{0}$	$\bar{0}$	$\bar{1}$	$\bar{2}$	$\bar{3}$
$\bar{1}$	$\bar{1}$	$\bar{2}$	$\bar{3}$	$\bar{0}$
$\bar{2}$	$\bar{2}$	$\bar{3}$	$\bar{0}$	$\bar{1}$
$\bar{3}$	$\bar{3}$	$\bar{0}$	$\bar{1}$	$\bar{2}$

$\times$	$\bar{0}$	$\bar{1}$	$\bar{2}$	$\bar{3}$
$\bar{0}$	$\bar{0}$	$\bar{0}$	$\bar{0}$	$\bar{0}$
$\bar{1}$	$\bar{0}$	$\bar{1}$	$\bar{2}$	$\bar{3}$
$\bar{2}$	$\bar{0}$	$\bar{2}$	$\bar{0}$	$\bar{2}$
$\bar{3}$	$\bar{0}$	$\bar{3}$	$\bar{2}$	$\bar{1}$

4.5.4 *Groups*

Definition 4.2. A **group** is a set G equipped with an operation $*$ satisfying the following axioms:

(1) **Closure of G under the operation** $*$. This axiom simply says that when we compose two elements of G, what we get is also an element of G: $x * y \in G$ for all $x, y \in G$.

(2) **Associativity of the operation** $*$. This property allows us to move the parenthesis freely when doing computations inside the group: $x * (y * z) = (x * y) * z$ for all $x, y, z \in G$.

(3) **Existence of an identity element.** There exists an element e (called the **identity element**) of G satisfying: $x * e = e * x = x$ for all $x \in G$.

(4) **Existence of inverses.** For every $x \in G$, there exists $y \in G$ such that $x * y = y * x = e$. The element $y \in G$ is called the **inverse** of x with respect to the operation $*$.

It is not hard to prove that the identity element of a group is unique. Also, if x is an element of a group, then the inverse of x is unique. If in addition to the above axioms, the operation $*$ is **commutative**, that is $x * y = y * x$ for all $x, y \in G$, then the group G is called **abelian**. A subset H of a group $(G, *)$ is called a **subgroup** of G if H is itself a group with respect to the same operation $*$.

It is convenient to use familiar notations for a group operation. The most familiar ones are of course $+$ and $\cdot$ (or just a concatenation). If we use the symbol $+$, we say that our group is *additive* and if the multiplication (or concatenation) is used, the group is called *multiplicative*. In an additive group, the identity element is called the *zero element* and denoted by 0 and the inverse of an element x of the group is called the *opposite* of x and denoted with $-x$. In a multiplicative group, the identity element is denoted by 1 and the inverse of an element x is denoted by x^{-1} or $\frac{1}{x}$.

Example 4.4. It should come as no surprise that the abstract definition of a group given above is a generalization of the well-known (additive) groups $(\mathbb{Z}, +)$ (the integers), $(\mathbb{Q}, +)$ (the rational numbers) and $(\mathbb{R}, +)$ (the real numbers). Note that $(\mathbb{Z}, +)$ is a subgroup of both $(\mathbb{Q}, +)$ and $(\mathbb{R}, +)$ and $(\mathbb{Q}, +)$ is a subgroup of $(\mathbb{R}, +)$. Changing the operation from addition to multiplication results in these sets losing their group structure: $(\mathbb{Z}, \times)$ is not a group because only ± 1 have their multiplicative inverses in $\mathbb{Z}$ and the inverse of any other integer is not an integer. $(\mathbb{Q}, \times)$ and $(\mathbb{R}, \times)$ are not groups since 0 does not have an inverse in these sets. However, and unlike $(\mathbb{Z}, \times)$, the sets $(\mathbb{Q}^*, \times)$ and $(\mathbb{R}^*, \times)$ are indeed groups where $\mathbb{Q}^*$ and $\mathbb{R}^*$ are respectively the sets of non-zero rational numbers and non-zero real numbers.

A group G is called *finite* if it contains a finite number of elements. In this case, we define the *order* of G, written $|G|$, as the number of elements in G. Finite groups play a pivotal role in many real life applications of mathematics. The following gives a classic example of a finite group.

Example 4.5. The set $\mathbb{Z}_n = \{\overline{0}, \overline{1}, \ldots, \overline{n-1}\}$ of integers modulo n defined in Section 4.5.3 above is an additive group for the addition modulo n.

All the group axioms can be easily verified. In particular, $\overline{0}$ is the zero element of the group and if $\overline{k} \in \mathbb{Z}_n$, then the opposite of $\overline{k}$ is $\overline{n-k}$ since $\overline{k} + \overline{n-k} = \overline{n} = \overline{0}$ in $\mathbb{Z}_n$.

What about the structure of $(\mathbb{Z}_n, \times)$ where $\times$ is the multiplication modulo n? Note first that the element $\overline{1} \in \mathbb{Z}_n$ is the identity element of $\mathbb{Z}_n$ for the multiplication modulo n since $\overline{k} \times \overline{1} = \overline{k \times 1} = \overline{k}$ for all $\overline{k} \in \mathbb{Z}_n$. But the element $\overline{0}$ has no multiplicative inverse since $\overline{k} \times \overline{0} = \overline{0} \neq \overline{1}$ for all $\overline{k} \in \mathbb{Z}_n$. This means that $(\mathbb{Z}_n, \times)$ is not a group. What about removing $\overline{0}$ from $\mathbb{Z}_n$ like we did for $\mathbb{Q}$ and $\mathbb{R}$? Would the resulting structure $(\mathbb{Z}_n^*, \times)$ be a group (like in the case of $(\mathbb{Q}^*, \times)$ and $(\mathbb{R}^*, \times)$)? A closer look at the multiplication table of $\mathbb{Z}_4$ given in Example 4.3 above quickly answers that question negatively: the element $\overline{2} \in \mathbb{Z}_4$ has no inverse since the row of $\overline{2}$ in that table does not contain $\overline{1}$. This is clearly not the case of the multiplication table of $\mathbb{Z}_3$ where every non-zero element seems to have an inverse, making $(\mathbb{Z}_3^*, \times)$ a group.

So under what conditions on n the set $(\mathbb{Z}_n^*, \times)$ becomes a (multiplicative) group? Part of the answer comes from the following observation: if n has a proper divisor d (so d divides n and $2 \leq d \leq n-1$), then the element $\overline{d}$ of $\mathbb{Z}_n$ does not have a multiplicative inverse. To see this, write $n = kd$ with $2 \leq k \leq n-1$. If $\overline{d}$ has a multiplicative inverse $\overline{d'}$, then

$$\overline{k} \times \overline{d} \times \overline{d'} = \underbrace{(\overline{k} \times \overline{d})}_{=\overline{n}=\overline{0}} \times \overline{d'} = \overline{0}.$$

On the other hand

$$\overline{k} \times \overline{d} \times \overline{d'} = \overline{k} \times \underbrace{(\overline{d} \times \overline{d'})}_{\overline{1}} = \overline{k} \neq \overline{0}$$

which leads to a contradiction. This shows that $(\mathbb{Z}_n^*, \times)$ is not a group if n has a proper divisor (since in that case, $\overline{d}$ has no multiplicative inverse in $\mathbb{Z}_n$). Integers with no proper divisors are called **prime integers**. For instance, $2, 3, 5, 7, 11$ are all prime. It is then natural to expect that if p is a prime integer, the set $\mathbb{Z}_p^* = \{\overline{1}, \overline{2}, \ldots, \overline{p-1}\}$ (of $p-1$ elements) is indeed a group for the multiplication modulo p. The proof of this fact uses some properties of the GCD (Greatest Common Divisor) of two integers that we will not include here but we state the result for future reference.

Theorem 4.4. If p is a prime integer, then the set $\mathbb{Z}_p^* = \{\overline{1}, \overline{2}, \ldots, \overline{p-1}\}$ (of $p-1$ elements) is a group for the multiplication modulo p.

Hence, $(\mathbb{Z}_2^*, \times)$, $(\mathbb{Z}_3^*, \times)$, $(\mathbb{Z}_5^*, \times)$ and $(\mathbb{Z}_{31}^*, \times)$ are all examples of multiplicative groups.

From this point on, and unless otherwise specified, the operation of a multiplicative group is simply denoted with a concatenation of elements and the identity element of the group is denoted simply by 1.

Definition 4.3. Let G be a (multiplicative) group, $g \in G$ and $m \in \mathbb{Z}$. If $m > 0$, we define g^m to be g composed with itself m times, that is

$$g^m = \underbrace{gg \cdots g}_{m \text{ times}}.$$

If $m < 0$, we define g^m to be $\left(g^{-1}\right)^{-m}$. This is well defined since in a group, every element has an inverse and $-m$ is now positive. If $m = 0$, we define g^m to be the identity element 1 of the group G.

Remark 4.4. In an additive group $(G, +)$, the notion of an "exponent" (or a "power") g^m of g translates to $g + g + \cdots + g = mg$.

The **exponent laws** for real numbers apply to elements of any group. Given a group G and elements g, h in G, then for any integers m, n we have

- $g^{m+n} = g^m g^n$
- $(g^m)^n = g^{mn}$
- If G is abelian, then $(gh)^m = g^m h^m$.

An important theorem in the theory of finite groups (due to Lagrange) relates the size of the subgroup to the size of the group.

Theorem 4.5. (Lagrange) If G is a finite group and H is a subgroup of G, then $|H|$ is a divisor of $|G|$.

Proof. Given $x \in G$, define xH as being the subset $\{xg; \ g \in H\}$ of G. Note that there are as many elements in xH as there are in H. To see this, let $g \neq g' \in H$ and suppose that $xg = xg'$. Since x^{-1} exists in G, multiplying both sides with x^{-1} yields $g = g'$ which is a contradiction. So, if $g \neq g'$, then $xg \neq xg'$ and so xH and H have the same number of elements. Note also that since H is a subgroup of G, $xH = H$ for any $x \in H$ (the operation is internal in H). Next, let $g \neq g' \in G$ and suppose that the sets gH and $g'H$ have an element $z \in G$ in common. Then there exist $h, h' \in H$ such that $z = gh = g'h'$ and we can write $g = g'h'h^{-1}$ (by multiplying both sides of $gh = g'h'$ with h^{-1} on the right). If $y \in gH$, then $y = gh''$

for some $h'' \in H$ and therefore $y = g'h'h^{-1}h''$. But $h'h^{-1}h'' \in H$ since H is a subgroup, so $y = g'h'h^{-1}h'' \in g'H$. This shows that gH is a subset of $g'H$. Similarly, we can show that $g'H$ is a subset of gH and conclude that $gH = g'H$. So as soon as the sets gH and $g'H$ have an element in common, they must be equal. In other words, the sets gH and $g'H$ are either disjoint (no elements in common) or they are the same set. Note also that $1H$ is simply the subgroup H. Finally, if $g \in G$, then $g = g1 \in gH$ since $1 \in H$. The group G can then be written as the union of pairwise disjoint subsets of the form:

$$G = H \cup g_1 H \cup \cdots \cup g_r H$$

with $|H| = |g_1 H| = \cdots = |g_r H|$. Thus, $|G| = |H| + |g_1 H| + \cdots + |g_r H| = (r+1)|H|$. We conclude that $|H|$ is a divisor of $|G|$. $\qquad\square$

Groups like $(\mathbb{Z}, +)$ and $(\mathbb{Z}_n, +)$ can be "generated" by a single element. For example, in the additive group $(\mathbb{Z}, +)$ every integer k can be generated using the element 1: $k = 1 + 1 + \cdots + 1 = k \times 1$. We say in this case that the group $(\mathbb{Z}, +)$ is **generated** by 1. Note also that -1 is a generator of $(\mathbb{Z}, +)$. In general, we have the following.

Definition 4.4. A group G is called **cyclic** if there exists an element $g \in G$ such that $G = \{g^m; \ m \in \mathbb{Z}\}$. In other words, every element of a cyclic group G can be written as a power of a fixed element $g \in G$. We say in this case that g is a **generator** of G and we write $G = \langle g \rangle$.

Example 4.6. The group $(\mathbb{Z}_5^*, \times) = \{\overline{1}, \overline{2}, \overline{3}, \overline{4}\}$ is cyclic with $\overline{2}$ is a generator since every element of the group can be expressed as a power of $\overline{2}$ as follows: $\overline{1} = \overline{2}^0$, $\overline{2} = \overline{2}^1$, $\overline{3} = \overline{2}^3$ and $\overline{4} = \overline{2}^2$.

Remark 4.5. By the exponent laws of a group, a cyclic group is always abelian.

Given a finite group G of order $n \geq 2$ and identity element 1, the exponent laws of G show in particular that the set $H_g = \{g^k; \ k \in \mathbb{Z}\}$ forms a subgroup of G for any $g \in G$. The subgroup H_g is called the *cyclic subgroup of G generated by g*. Since G is finite, we can find $k, m \in \mathbb{Z}$ with $k < m$ and $g^k = g^m$ (otherwise H_g would be infinite). Multiplying both sides of $g^k = g^m$ with g^{-k} gives that $g^{m-k} = 1$. So the set $P_g = \{l \in \mathbb{Z}; \ l > 0 \text{ and } g^l = 1\}$ is not empty. The **order** of the element g, denoted by $|g|$, is defined as being the smallest element of P_g. That is, $|g|$ is the smallest positive integer l satisfying $g^l = 1$. Therefore, the subgroup

H_g is equal to $\{g^0 = 1, g, g^2, \ldots, g^{|g|-1}\}$ and the order of $g \in G$ is equal to the order of the subgroup H_g generated by g.

Theorem 4.6. If G is a finite group of order n, then $g^n = 1$ for any $g \in G$.

Proof. By Lagrange Theorem, we know that $|g| = |H_g|$ is a divisor of n. Write $n = k|g|$ for some $k \in \mathbb{N}$, then $g^n = g^{k|g|} = \left(g^{|g|}\right)^k = 1^k = 1$ since $g^{|g|} = 1$ by definition of the order of g. $\qquad\square$

4.5.5 *Fields - An introduction and basic results*

We have seen that the sets $(\mathbb{Q}, +)$, $(\mathbb{R}, +)$ and $(\mathbb{Z}_n, +)$ are all examples of additive groups. There is however more to say about their structures. Each one of these sets is also equipped with a multiplication that interacts well with the addition to give each of them a more "advanced" structure known as a **field**. The additive group $(\mathbb{Z}, +)$ is also equipped a multiplication but its structure differs from that of $\mathbb{Q}$ and $\mathbb{R}$ in the following key property: the inverse of a non-zero rational number (respectively a non-zero real number) is also a rational number (respectively a real number) while the inverse of an integer is not an integer, except for ± 1.

Field theory has deep roots in the history of abstract algebra and has been perceived for many years as a purely academic topic in mathematics. But with the increasing demand on improving the technology and security of communication, field theory started to play a central role in many real life applications. In what follows we give a brief introduction and some important facts about this theory, enough to be able to prove our main result (Theorem 4.1).

Definition 4.5. A **field** is a nonempty set $\mathbb{F}$ together with two internal operations often called an addition and a multiplication (denoted as usual by $+$ and $\times$ or just a concatenation respectively) such that the following axioms are satisfied.

- $(\mathbb{F}, +)$ is an abelian group with identity element denoted by 0;
- $(\mathbb{F}^*, \times)$ is an abelian group where $\mathbb{F}^* = \{x \in \mathbb{F}; x \neq 0\}$;
- The multiplication is distributive over addition: $x(y + z) = xy + xz$ for all $x, y, z \in \mathbb{F}$.

In what follows, 0 and 1 denote the identity elements of the groups $(\mathbb{F}, +)$ and $(\mathbb{F}^*, \times)$ respectively for the field $(\mathbb{F}, +, \times)$. The first one is referred to as the zero element and the second as the identity element of the field. There is only one field, referred to as the **zero field**, where the zero element and the identity element are the same. This is a set with only one element 0 with the obvious rules: $0 + 0 = 0 \times 0 = 0$. Any other field is called a **non-zero field**.

The set $(\mathbb{Z}, +, \times)$ is not a field since $(\mathbb{Z}^*, \times)$ is not a multiplicative group. The sets $\mathbb{Q}$, $\mathbb{R}$ and $\mathbb{C}$ (of rational numbers, real numbers and complex numbers respectively) with the usual addition and multiplication of numbers are classic examples of a field structure. However, these are not the kind of fields used in real applications. In what follows we look at fields containing a finite number of elements that we call **finite fields**.

4.5.6 The field $\mathbb{Z}_p$

The multiplication table of $\mathbb{Z}_4$ given in Example 4.3 above reveals the following surprising fact: $\overline{2} \times \overline{2} = \overline{0}$ in spite of the fact that $\overline{2} \neq \overline{0}$. This cannot happen in a field as shown in the following proposition.

Proposition 4.1. Let $\mathbb{F}$ be a non-zero field. Then

(1) $a0 = 0$ for all $a \in \mathbb{F}$.
(2) If $a, b \in \mathbb{F}$ are such that $ab = 0$, then either $a = 0$ or $b = 0$.

Proof.

(1) $a0 = a(0 + 0) = a0 + a0$ (by the distributivity property of a field). As an element of a field, $a0$ must have an additive inverse (or opposite) $-a0$. Adding $-a0$ to the last equation gives $0 = a0$.
(2) Assume $ab = 0$. If $a \neq 0$, then a admits a multiplicative inverse a^{-1} since $(\mathbb{F}^*, \times)$ is a group. Multiplying both sides of $ab = 0$ with a^{-1} gives

$$a^{-1}(ab) = a^{-1}0 \Rightarrow \underbrace{(a^{-1}a)}_{1}b = 0 \Rightarrow 1b = 0 \Rightarrow b = 0.$$

We conclude that at least one of the elements a, b must be zero. $\square$

The proposition above shows in particular that $\mathbb{Z}_4$, equipped with the addition and the multiplication modulo 4, is not a field since $\overline{2} \times \overline{2} = \overline{0}$ and $\overline{2} \neq \overline{0}$. Similarly, $\overline{2} \times \overline{3} = \overline{6} = \overline{0}$ in $\mathbb{Z}_6$ with both $\overline{2}, \overline{3}$ non-zero. On the other hand, addition and multiplication tables of $\mathbb{Z}_3$ show that $\mathbb{Z}_3$ is indeed a field. The fact that 4 and 6 can be written as 2×2 and 2×3 respectively is the main reason why $(\mathbb{Z}_4, +, \times)$ and $(\mathbb{Z}_6, +, \times)$ are not fields.

In general, if n is not a prime integer, then n can be written under the form $n = pq$ where $1 < p, q < n$. This translates in $\mathbb{Z}_n$ into the equation $\bar{p} \times \bar{q} = \bar{n} = \bar{0}$ with both $\bar{p}, \bar{q}$ non-zero. This means that $\mathbb{Z}_n$ is not a field. On the other hand, Theorem 4.4 above shows that $\mathbb{Z}_p^*$ is a (multiplicative) group if p is a prime integer. We then have the following result.

Theorem 4.7. $\mathbb{Z}_p$ is a field (for the addition and a multiplication modulo p) if and only if p is a prime integer.

Hence, $\mathbb{Z}_2$, $\mathbb{Z}_5$ and $\mathbb{Z}_7$ are all examples of finite fields.

Remark 4.6. It can be shown (but we will not show it here) that any finite field $\mathbb{F}$ containing p elements where p is a prime integer is actually a "copy" of $\mathbb{Z}_p$ (formally, we say $\mathbb{F}$ is *isomorphic* to $\mathbb{Z}_p$). By a "copy", we mean that we can relabel the elements of $\mathbb{F}$ to match those of $\mathbb{Z}_p$ (namely, $\bar{1}, \bar{2}, \ldots,$ $\overline{p-1}$) in such a way the addition and multiplication tables of $\mathbb{F}$ are the same as those of $\mathbb{Z}_p$. In other words, there is a unique field containing p elements for each prime integer p. This field is denoted by $\mathbb{F}_p$.

From this point on, we will omit the "over line" in expressing the element $\bar{a}$ of $\mathbb{Z}_p$ and just write a for simplicity. For instance, we write $\mathbb{Z}_3 = \{0, 1, 2\}$ and $\mathbb{Z}_5 = \{0, 1, 2, 3, 4\}$.

The field $\mathbb{Z}_p$ (or $\mathbb{F}_p$) is just a particular example of a more general family of finite fields. The following is a key result in the theory of finite fields. We omit the proof as it is beyond the scope of this book.

Theorem 4.8. The number of elements in any non-zero finite field is p^r for some prime integer p and positive integer r. Conversely, given a prime integer p and a positive integer r, there exists a unique finite field (uniqueness up to relabeling the elements) containing p^r elements, that we denote by $\mathbb{F}_{p^r}$.

The field $\mathbb{F}_{p^r}$ plays a key role in understanding the properties of the sequence produced by a LFSR. Our next task is to shed more light on its structure. For this, we need the notion of a polynomial over a field.

4.5.7 *Polynomials over a field*

In all what follows, $\mathbb{F}$ denotes an arbitrary non-zero field (not necessarily finite), p a prime integer and r a positive integer. We will "cook" the field $\mathbb{F}_{p^r}$ following two recipes. The main ingredient in both recipes is the notion of polynomials with coefficients in the field $\mathbb{F}$. These are the same type of polynomials that you always dealt with except that the coefficients are no longer restricted to real numbers.

Definition 4.6. A **polynomial** in one variable x over $\mathbb{F}$ is an expression of the form

$$p(x) = a_n x^n + a_{n-1} x^{n-1} + \cdots + a_1 x + a_0$$

where $a_i \in \mathbb{F}$ for each $i \in \{0, 1, \ldots, n\}$. If $a_n \neq 0$ (with 0 being the zero element of the field $\mathbb{F}$), then we say that $p(x)$ is of **degree** n and we write $\deg p(x) = n$. In this case, the coefficient a_n is called the **leading coefficient** of $p(x)$. A **monic polynomial** is a polynomial with leading coefficient 1 (the identity element of the field $\mathbb{F}$). If $a_i = 0$ for all i, we say that $p(x)$ is the **zero polynomial**. The degree of the zero polynomial is defined to be $-\infty$. Note that any element of the field $\mathbb{F}$ can be considered as a polynomial of degree 0 that we usually call a **constant polynomial**. The set of all polynomials in one variable x over $\mathbb{F}$ is denoted by $\mathbb{F}[x]$.

We define addition and multiplication in $\mathbb{F}[x]$ in the usual way of adding and multiplying two polynomials with real coefficients with the understanding that the involved operations on the coefficients are done in the field $\mathbb{F}$. These two operations inside $\mathbb{F}[x]$ do not give this set the status of a field since, for example, the multiplicative inverse of the polynomial $x \in \mathbb{F}[x]$ does not exist (no polynomial $p(x)$ exists such that $x p(x) = 1$).

Remark 4.7. We are mainly interested in polynomials over the finite fields $\mathbb{Z}_p$ (for prime p) and one has to be careful when computing modulo the prime p. For instance, let $p(x) = x^2 + x + 1$ and $q(x) = x + 1$ considered as polynomials in $\mathbb{Z}_2[x]$. Then $p(x) + q(x) = x^2 + 2x + 2 = x^2$ since in the field $\mathbb{Z}_2$, $2 = 0$ (remember: the coefficient 2 here means $\bar{2}$). Also $p(x)q(x) = x^3 + 2x^2 + 2x + 1 = x^3 + 1$ for the same reason. Now, consider the same polynomials $p(x) = x^2 + x + 1$ and $q(x) = x + 1$ but this time as elements of $\mathbb{Z}_3[x]$. Then $p(x) + q(x) = x^2 + 2x + 2$ and $p(x)q(x) = x^3 + 2x^2 + 2x + 1$.

Similar to integers, we can talk about division of polynomials and we have a division algorithm in $\mathbb{F}[x]$.

Definition 4.7. Let $p(x)$, $q(x)$ be two polynomials in $\mathbb{F}[x]$ with $p(x)$ not equal to the zero polynomial. We say that $p(x)$ is a *divisor* of $q(x)$ (or that $p(x)$ *divides* $q(x)$) if $q(x) = p(x)k(x)$ for some $k(x) \in \mathbb{F}[x]$. In this case, we also say that $q(x)$ is a multiple of $p(x)$.

Example 4.7. In $\mathbb{Z}_2[x]$, $p(x) = x^2 + x + 1$ is a divisor of $x^3 + 1$ since $(x+1)(x^2 + x + 1) = x^3 + 1$ (see Remark 4.7 above).

Example 4.8. $x^4 - 1$ is a multiple of $x^2 + 1$ in $\mathbb{F}[x]$ for any field $\mathbb{F}$ since $x^4 - 1 = (x^2 - 1)(x^2 + 1)$ holds regardless of the base field $\mathbb{F}$.

Division algorithm of $\mathbb{F}[x]$. Given two polynomials $f(x)$ and $g(x)$ in $\mathbb{F}[x]$ with $\deg g(x) \geq 1$, then uniquely determined polynomials $q(x)$ and $r(x)$ exist in $\mathbb{F}[x]$ such that

(1) $f(x) = g(x)q(x) + r(x)$;
(2) Either $r(x)$ is the zero polynomial or $\deg r(x) < \deg g(x)$.

The polynomial $q(x)$ is called the *quotient* of the division and $r(x)$ is called the *remainder*. Note that if $\deg f(x) < \deg g(x)$, then we can write $f(x) = g(x) \cdot 0 + f(x)$ with 0 as quotient and $f(x)$ as remainder. As in the case of integers, the above algorithm guarantees the existence of a quotient and a remainder but it does not say much about how to find them. Usually the long division of polynomials is used to that end.

Example 4.9. Let $p(x) = x^4 + 2x^3 + x + 2$ and $k(x) = x^2 + x + 1$ considered as polynomials in $\mathbb{Z}_3[x]$ where as usual $\mathbb{Z}_3 = \{0, 1, 2\}$. We can perform the long division of $p(x)$ by $k(x)$ the usual way but bear in mind that we are not dealing with real numbers here but rather elements of the field $\mathbb{Z}_3$.

$$
\begin{array}{r}
x^2 + x - 2 \\
x^2 + x + 1 \overline{)\; x^4 + 2x^3 \qquad\quad + x + 2} \\
\underline{-x^4 - x^3 - x^2} \\
x^3 - x^2 + x \\
\underline{-x^3 - x^2 - x} \\
-2x^2 \qquad + 2 \\
\underline{2x^2 + 2x + 2} \\
2x + 4
\end{array}
$$

The quotient is $q(x) = x^2 + x - 2 = x^2 + x + 1$ (since $-2 = 1$ in the field $\mathbb{Z}_3$) and the remainder is $r(x) = 2x + 4 = 2x + 1$ (since $4 = 1$ in the field $\mathbb{Z}_3$).

We give next an analogue of prime integers for polynomials.

Definition 4.8. A non-zero polynomial $p(x) \in \mathbb{F}[x]$ is called **irreducible over** $\mathbb{F}$ if it cannot be written as the product of two non-constant polynomials in $\mathbb{F}[x]$. In other words, $p(x)$ is irreducible if and only if the only way an equality of the form $p(x) = p_1(x)p_2(x)$ with $p_1(x)$, $p_2(x) \in \mathbb{F}[x]$ can occur is when either $p_1(x)$ or $p_2(x)$ is a constant polynomial. Consequently, if $p(x)$ is irreducible of degree r, then it does have a non-constant polynomial divisor (or factor) of degree strictly less than r.

The notion of irreducibility for polynomials depends largely on the coefficient field. If $\mathbb{F}_1$ is a field contained in a larger field $\mathbb{F}_2$, it could very well happen that a polynomial $p(x)$ is irreducible as an element of $\mathbb{F}_1[x]$ but not as an element of $\mathbb{F}_2[x]$.

Example 4.10. The polynomial $p(x) = x^2 - 2$ is irreducible as element of $\mathbb{Q}[x]$ but not as an element of $\mathbb{R}[x]$ since $p(x) = (x - \sqrt{2})(x + \sqrt{2})$ and each one of the polynomials $(x - \sqrt{2})$, $(x + \sqrt{2})$ is non-constant in $\mathbb{R}[x]$.

More interesting examples arise in the case of finite fields.

Example 4.11. The polynomial $p(x) = x^2 + 1$ is not irreducible over $\mathbb{Z}_2$ since $(x + 1)(x + 1) = x^2 + 2x + 1 = x^2 + 1$ in $\mathbb{Z}_2[x]$. Note that $x^2 + 1$ is clearly irreducible in $\mathbb{R}[x]$.

Remark 4.8. It can be shown that if $\mathbb{F}$ is a finite field, then there exists a monic irreducible polynomial of degree r in $\mathbb{F}[x]$ for any positive integer r.

Similar to the arithmetics "modulo n" in $\mathbb{Z}$, we can define operations "modulo $p(x)$" in $\mathbb{F}[x]$ for some fixed polynomial $p(x) \in \mathbb{F}[x]$. First, a definition.

Definition 4.9. Let $\mathbb{F}$ be a field, $p(x) \in \mathbb{F}[x]$ a non-zero polynomial. We say that the two polynomials $f(x), g(x) \in \mathbb{F}[x]$ are **congruent modulo** $p(x)$, and we write $f(x) \equiv g(x) \pmod{p(x)}$ (or sometimes $f(x) = g(x) \pmod{p(x)}$), if $p(x)$ divides the difference $f(x) - g(x)$. Note that (like in the case of integers) the fact that $p(x)$ divides $f(x) - g(x)$ is equivalent to $f(x)$ and $g(x)$ having the same remainder when divided with $p(x)$.

Example 4.12. $x^3 + 2x^2 - 1 \equiv x^2 - 1 \pmod{x + 1}$ in $\mathbb{R}[x]$ since $x^3 + 2x^2 - 1 - (x^2 - 1) = x^3 + x^2 = x^2(x + 1)$.

Example 4.13. $x^3 + 3x \equiv x^3 - x^2 - 2x - 1 \pmod{x^2 + 1}$ in $\mathbb{Z}_5[x]$ since $x^3 + 3x - (x^3 - x^2 - 2x - 1) = x^2 + 5x + 1 = x^2 + 1$ (remember that $5 = 0$ in $\mathbb{Z}_5$).

The division algorithm is at the heart of arithmetic modulo $p(x)$ in $\mathbb{F}[x]$. If $f(x) = p(x)q(x) + r(x)$, then $f(x) - r(x) = p(x)q(x)$ and consequently, $f(x) \equiv r(x) \pmod{p(x)}$. Like in the case of integers, given a non-zero polynomial $p(x) \in \mathbb{F}[x]$, we can group the polynomials of $\mathbb{F}[x]$ in "classes" according to their remainder upon division by $p(x)$. So two polynomials $f(x)$ and $g(x)$ are "equal" modulo $p(x)$ if and only if they belong to the same class, or equivalently they have the same remainder when divided by $p(x)$.

For a non-zero polynomial $p(x) \in \mathbb{F}[x]$, we denote by $\mathbb{F}[x]/\langle p(x) \rangle$ the set of all classes of $\mathbb{F}[x]$ modulo $p(x)$. In other words, $\mathbb{F}[x]/\langle p(x) \rangle$ is the set of all possible remainders upon (long) division with the polynomial $p(x)$. Like in the case of integers modulo n, addition and multiplication (modulo $p(x)$) in $\mathbb{F}[x]/\langle p(x) \rangle$ are well defined operations in the sense that they do not depend on the representatives of the classes.

Remark 4.9. If $p(x)$ is a non-zero polynomial in $\mathbb{F}[x]$ and α is a non-zero element of $\mathbb{F}$, then we can easily verify that $f(x) \equiv g(x) \pmod{p(x)}$ if and only if $f(x) \equiv g(x) \pmod{\alpha p(x)}$ for any polynomials $f(x)$ and $g(x)$ in $\mathbb{F}[x]$. As a consequence, if $p(x) = a_n x^n + \cdots + a_1 x + a_0 \in \mathbb{F}[x]$ then the arithmetic modulo $p(x)$ is the same as the arithmetic modulo $a_n^{-1} p(x) = x^n + \cdots + a_n^{-1} a_1 x + a_n^{-1} a_0$. This allows us to assume, without any loss of generality, that the polynomial $p(x)$ is monic when looking at the structure of $\mathbb{F}[x]/\langle p(x) \rangle$.

Example 4.14. Let $p(x) = x^2 + 2x \in \mathbb{Z}_3[x]$. Let us see how we can add and multiply the two polynomials $h(x) = x^3 + x^2$ and $k(x) = x^2 + 2x + 2$ of $\mathbb{Z}_3[x]$ modulo $p(x)$. First note that $h(x) + k(x) = x^3 + 2x^2 + 2x + 2$ and $h(x)k(x) = x^5 + 3x^4 + 4x^3 + 2x^2 = x^5 + x^3 + 2x^2$ (since $3 = 0$ and $4 = 1$ in $\mathbb{Z}_3$). We start by performing the long division of both $h(x) + k(x)$ and $h(x)k(x)$ by $p(x)$. This leads to the following two relations (the reader is encouraged to do the long division):

$$h(x) + k(x) = xp(x) + (2x + 2),$$

$$h(x)k(x) = \left(x^3 - 2x^2 + 5x - 8\right) p(x) + 16x$$
$$= \left(x^3 - 2x^2 + 5x - 8\right) p(x) + x \quad \text{since } 16 = 1 \text{ in } \mathbb{Z}_3.$$

We conclude that $h(x) + k(x) = 2x + 2 \pmod{x^2 + 2x}$ and $h(x)k(x) = x \pmod{x^2 + 2x}$.

If $p(x) \in \mathbb{F}[x]$ is not irreducible over $\mathbb{F}$, we would have an equation of type $hq = 0$ in the set $\mathbb{F}[x]/\langle p(x)\rangle$ with at least one of h or q is non-zero (can you see why?). This would deprive $\mathbb{F}[x]/\langle p(x)\rangle$ from having a field structure with respect to addition and multiplication mod $p(x)$ by Proposition 4.1 above. This suggests that $\mathbb{F}[x]/\langle p(x)\rangle$ (with operations modulo $p(x)$) is a field only in the case where $p(x)$ is an irreducible polynomial. To completely prove that fact, one would need the notion of greatest common divisor of two polynomials and the Euclidian algorithm to find it. These are techniques that interested reader can pick up from any abstract algebra book. We state this fact for future reference.

Theorem 4.9. Let $p(x) \in \mathbb{F}[x]$ be a non-constant polynomial. The set $\mathbb{F}[x]/\langle p(x)\rangle$ equipped with addition and multiplication modulo $p(x)$ is a field if and only if $p(x)$ is an irreducible polynomial.

4.5.8 The field $\mathbb{F}_{p^r}$ - A first approach

Given a prime number p and a positive integer r, we are now ready to provide a first approach to construct the field $\mathbb{F}_{p^r}$ containing p^r elements. Start by choosing an irreducible and monic polynomial $p(x)$ of degree r with coefficients in the finite field $\mathbb{F}_p = \{0, 1, \ldots, p-1\}$. Remark 4.8 above guarantees the existence of such a polynomial. The irreducibility of $p(x)$ gives $\mathbb{F}_p[x]/\langle p(x)\rangle$ the structure of a field by Theorem 4.9. Since $\mathbb{F}_p[x]/\langle p(x)\rangle$ consists of all remainders upon division by $p(x)$ and since any remainder is of degree at most $r - 1$, elements of this field are of the form

$$c_0 + c_1 t + c + 2t^2 + \cdots + c_{r-1}t^{r-1}$$

with $c_i \in \mathbb{F}$ for each i. Such a polynomial has r coefficients each of which can take on p values in the field $\mathbb{F}_p$. This means that there are exactly p^r elements in the field $\mathbb{F}_p[x]/\langle p(x)\rangle$. The following example shows how we can explicitly find the elements of $\mathbb{F}_p[x]/\langle p(x)\rangle$ and more importantly how to perform the field operations on these elements.

Example 4.15. Consider the polynomial $p(x) = x^3 + x + 1$ of $\mathbb{F}_2[x]$. We start by proving that $p(x)$ is irreducible over $\mathbb{F}_2$. Suppose not, then the polynomial can factor into a product of two non-constant polynomials in $\mathbb{F}_2[x]$. Since $p(x)$ is of degree 3, at least one of the factors must be linear (of degree 1). So there exist $a, b, c \in \mathbb{Z}_2$ such that $x^3 + x + 1 = (x+a)(x^2 + bx + c)$. In particular, $x = -a$ is a root of the polynomial in $\mathbb{F}_2$. But no such root exists in $\mathbb{F}_2$ since $p(0) = 1 \neq 0$ and $p(1) = 3 = 1 \neq 0$. We conclude that $p(x) = x^3 + x^2 + 1$ is irreducible and hence $\mathbb{Z}_2[x]/\langle x^3 + x + 1 \rangle$ is indeed a field. Next, we look more closely at the list the elements of this field. We know that

$$\mathbb{Z}_2[x]/\langle x^3 + x + 1 \rangle = \left\{ a_0 + a_1 t + a_2 t^2;\ a_0, a_1, a_2 \in \mathbb{Z}_2 \right\}.$$

There are exactly $2^3 = 8$ elements in this field, namely:

$$\mathbb{Z}_2[x]/\langle x^3 + x + 1 \rangle$$
$$= \left\{ 0,\ 1,\ 1 + t + t^2,\ 1 + t,\ 1 + t^2,\ t + t^2,\ t,\ t^2 \right\}. \tag{4.7}$$

It is important to know is how to actually perform operations on the elements inside the field. Note first that $x^3 + x + 1 \equiv 0 \pmod{x^3 + x + 1}$ which corresponds to the relation $t^3 + t + 1 = 0$ in $\mathbb{Z}_2[x]/\langle x^3 + x + 1 \rangle$. As it turns out, this relation is the "vehicle" that brings any multiplication $\alpha\beta$ of elements of $\mathbb{Z}_2[x]/\langle x^3 + x + 1 \rangle$ to one element in the set (4.7) above. For example, when the elements $1 + t + t^2$ and t^2 are multiplied together, we find $(1 + t + t^2)(t^2) = t^2 + t^3 + t^4$ which does not appear as one of the elements of the field (4.7). But since $t^3 + t + 1 = 0$, we get that $t^3 = -t - 1 = t + 1$ ($-\alpha = \alpha$ in $\mathbb{Z}_2$ since $\alpha + \alpha = 2\alpha = 0$ for any element α) and so

$$(1 + t + t^2)(t^2) = t^2 + t^3 + t^4$$
$$= t^2 + (t+1) + t(t+1) = t^2 + (t+1)^2 = 2t^2 + 2t + 1 = 1.$$

Likewise,

$$(1 + t)(1 + t + t^2) = 1 + t + t^2 + t + t^2 + t^3$$
$$= 1 + 2t + 2t^2 + t^3 = 1 + t^3 = 1 + (-t - 1) = -t = t.$$

Another important feature one should notice about the multiplication in $\mathbb{Z}_2[x]/\langle x^3 + x + 1 \rangle$ is the fact that every non-zero element of this field can be expressed as a power of the element $\alpha = t$ of the field. Here is why: $\alpha^0 = 1$, $\alpha^1 = t$, $\alpha^2 = t^2$, $\alpha^3 = t + 1$, $\alpha^4 = t^2 + t$, $\alpha^5 = 1 + t + t^2$, $\alpha^6 = 1 + t^2$, $\alpha^7 = 1$.

This is not a coincidence according to the following proposition.

Proposition 4.2. If $(\mathbb{F}, +, \times)$ is a finite non-zero field, then $(\mathbb{F}^*, \times)$ is a cyclic group. Here $\mathbb{F}^*$ is, as usual, the set $\mathbb{F}$ from which the zero element of the field is removed.

Proof.   We .give a sketch of the proof leaving it to the reader to fill in some details. By Theorem 4.8 above, the field $\mathbb{F}$ contains p^r elements for some prime number p and positive integer r. So $\mathbb{F}^*$ has $l = p^r - 1$ elements. Write l as a product of powers of primes $l = p_1^{q_1} p_2^{q_2} \cdots p_c^{q_c}$ (called prime decomposition) where the p_i's are distinct prime numbers and the q_i's are positive integers. If we can find $\alpha \in \mathbb{F}^*$ with order l, then we are done since this would mean that the cyclic subgroup $\langle \alpha \rangle$ has the same number of elements as the full group $\mathbb{F}^*$. Let $l_i = \frac{l}{p_i}$ for $i = 1, \ldots, c$, then the polynomial $x^{l_i} - 1$ of degree l_i cannot have every element of $\mathbb{F}^*$ as a root since $l_i < l$ and the polynomial can have at most l_i roots. For each $i = 1, \ldots, c$, we can then choose $\xi_i \in \mathbb{F}^*$ such that $\xi_i^{l_i} - 1 \neq 0$. Let $\alpha_i = \xi_i^{\frac{l}{p_i^{q_i}}}$, then the order $|\alpha_i|$ of α_i is a divisor of $p_i^{q_i}$ since $(\alpha_i)^{p_i^{q_i}} = \xi_i^l = 1$. If $|\alpha_i| \neq p_i^{q_i}$, then $|\alpha_i| = p_i^m$ for some $m < q_i$ and so $|\alpha_i|$ is a divisor of $p_i^{q_i-1}$. But if that is true, then $(\alpha_i)^{p_i^{q_i-1}} = 1$ which implies that $\xi_i^{l_i} = 1$, a contradiction to the fact that $\xi_i^{l_i} - 1 \neq 0$. This shows that the order of α_i is $p_i^{q_i}$. Now let $\alpha = \alpha_1 \alpha_2 \cdots \alpha_c \in \mathbb{F}^*$, then it can be shown that the order of α is the product of the orders of the elements α_i, namely l.   □

Definition 4.10.   A **primitive element** of a finite non-zero field $(\mathbb{F}, +, \times)$ is any generator of the cyclic group $(\mathbb{F}^*, \times)$. In other words, $\alpha \in \mathbb{F}^*$ is primitive if $\mathbb{F}^* = \{1, \alpha, \alpha^2, \ldots, \alpha^{r-2}\}$ where $r = |\mathbb{F}|$.

Example 4.16. In Example 4.15 above, we saw that every non-zero element of the field $\mathbb{Z}_2[x]/\langle x^3 + x + 1 \rangle$ is a power of $\alpha = t$. Thus, $\alpha = t$ is a primitive element of $\mathbb{Z}_2[x]/\langle x^3 + x + 1 \rangle$.

4.5.9   *The field $\mathbb{F}_{p^r}$ - A second approach*

Now for the second approach to construct $\mathbb{F}_{p^r}$. Recall that the field $\mathbb{F}_p$ containing p elements is nothing but a copy of the field $\mathbb{Z}_p$ of all integers modulo p.

Consider   the   set $\mathbb{Z}_p^r = \underbrace{\mathbb{Z}_p \times \mathbb{Z}_p \times \cdots \times \mathbb{Z}_p}_{r}$ of   all r-tuples

$(a_0, a_1, \ldots, a_{r-1})$ where $a_i \in \mathbb{Z}_p$ for all i. Since each component a_i can take p values, the set $\mathbb{Z}_p^r$ consists of p^r elements. We define an addition and a multiplication in $\mathbb{Z}_p^r$ as follows.

- The addition is defined on $\mathbb{Z}_p^r$ the natural way (componentwise):

$$(a_0, a_1, \ldots, a_{r-1}) + (b_0, b_1, \ldots, b_{r-1}) = (a_0 + b_0, a_1 + b_1, \ldots, a_{r-1} + b_{r-1})$$

  where $a_i + b_i$ represents the addition modulo p in $\mathbb{Z}_p$.

- Unlike addition, the multiplication on $\mathbb{Z}_p^r$ will probably appear to you as very "unnatural". We start by fixing an irreducible and monic polynomial $M(t) = t^r + m_{r-1}t^{r-1} + \cdots + m_1 t + m_0$ of degree r in $\mathbb{Z}_p[t]$ (by Remark 4.8 above, we know that such a polynomial exists). Each r-tuple $(a_0, a_1, \ldots, a_{r-1}) \in \mathbb{Z}_p^r$ is identified with the polynomial $p(t) = a_{r-1}t^{r-1} + \cdots + a_1 t + a_0 \in \mathbb{Z}_p[t]$ of degree less than or equal to $r - 1$ with coefficients in the field $\mathbb{Z}_p$. To define the multiplication of two r-tuples $(a_0, a_1, \ldots, a_{r-1})$ and $(b_0, b_1, \ldots, b_{r-1})$ of $\mathbb{Z}_p^r$, we start by writing the corresponding polynomials in $\mathbb{Z}_p[t]$:

$$p(t) = a_{r-1}t^{r-1} + \cdots + a_1 t + a_0, \quad q(t) = b_{r-1}t^{r-1} + \cdots + b_1 t + b_0,$$

  then we multiply the two polynomials together in the usual way by regrouping terms in $t^0, t, t^2, \ldots, t^{2(r-1)}$:

$$p(t)q(t) = a_{r-1}b_{r-1}t^{2(r-1)} + \cdots + (a_0 b_1 + a_1 b_0)t + a_0 b_0$$

  which in turns is congruent to its remainder $R(t)$ modulo $M(t)$ as an element of the field $\mathbb{Z}_p[t]/\langle M(t)\rangle$. Since the remainder is of degree less than or equal to $r - 1$, it can be written in the form $R(t) = \alpha_{r-1}t^{r-1} + \cdots + \alpha_1 t + \alpha_0$ where $\alpha_i \in \mathbb{F}$ for all i. Now define the multiplication of the two r-tuples $(a_0, a_1, \ldots, a_{r-1})$ and $(b_0, b_1, \ldots, b_{r-1})$ as being the r-tuple consisting of the coefficients of $R(t)$:

$$(a_0, a_1, \ldots, a_{r-1}) \times (b_0, b_1, \ldots, b_{r-1}) = (\alpha_0, \alpha_1, \ldots, \alpha_{r-1}).$$

One can verify that the set $\mathbb{Z}_p^r$ equipped with the above addition and multiplication with respect to a monic irreducible polynomial $M(t)$ is indeed a field.

Remark 4.10. The key feature in this second approach is the fact that it allows us to look at the r-tuples of $\mathbb{Z}_p^r$ as polynomials. The two fields $\mathbb{F}_p^r$ and $\mathbb{Z}_p[t]/\langle M(t)\rangle$ are copies of each others. Formally, we say that they are *isomorphic*.

Example 4.17. Consider the 3-tuples $(1, 0, 1)$ and $(1, 1, 1)$ as elements of $\mathbb{Z}_2^3$. As polynomials, these 3-tuples can be identified with $t^2 + 1$ and $t^2 + t + 1$ respectively. We have seen in Example 4.15 above that the polynomial $M(t) = t^3 + t + 1 \in \mathbb{Z}_2[t]$ is irreducible. Let us multiply the two 3-tuples with respect to $M(t)$:

$$(t^2 + 1)(t^2 + t + 1) = t^4 + t^3 + 2t^2 + t + 1 = t^4 + t^3 + t + 1$$

(remember that $2 = 0$ in $\mathbb{Z}_2$). Now we divide $t^4 + t^3 + t + 1$ with $t^3 + t + 1$:

$$
\begin{array}{r}
t + 1 \\
t^3 + t + 1) \overline{\quad t^4 + t^3 \qquad\quad + t + 1} \\
-t^4 \qquad -t^2 - t \\
\overline{\qquad t^3 - t^2 \qquad + 1} \\
-t^3 \qquad -t - 1 \\
\overline{\qquad\qquad -t^2 - t}
\end{array}
$$

and get a remainder of $-t^2 - t = t^2 + t$. The coefficients of this remainder are represented with the 3-tuple $(0, 1, 1)$. So, $(1, 0, 1) \times (1, 1, 1) = (0, 1, 1)$.

Definition 4.11. An irreducible monic polynomial $F(x) \in \mathbb{Z}_p[x]$ of degree r is called a **primitive polynomial** over $\mathbb{Z}_p$ if the monomial t is a primitive element of the field $\mathbb{Z}[x]/\langle F(x) \rangle$ when the elements of the field are identified with polynomials of the form $a_{r-1}t^{r-1} + \cdots + a_1 t + a_0$ with $a_i \in \mathbb{Z}_p$ for all i.

Example 4.18. In Example 4.15 above, the polynomial $P(x) = x^3 + x + 1 \in \mathbb{Z}_2[x]$ is primitive since it is irreducible, monic and t is a primitive element of the field $\mathbb{Z}_2[x]/\langle x^3 + x + 1 \rangle$.

Example 4.19. The polynomial $x^6 + x^3 + 1 \in \mathbb{Z}_2[x]$ is irreducible since it has no roots in $\mathbb{Z}_2$. In the field $\mathbb{Z}_2[x]/\langle x^6 + x^3 + 1 \rangle$, the equation $t^6 + t^3 + 1 = 0$ is equivalent to $t^6 = -t^3 - 1 = t^3 + 1$. This gives the following powers of the monomial t:

$$t^7 = t^4 + t, \; t^8 = t^5 + t^2, \; t^9 = t^6 + t^3 = t^3 + 1 + t^3 = 2t^3 + 1 = 1.$$

The fact that $t^9 = 1$ and that the multiplicative group of $\mathbb{F}[x]/\langle x^6 + x^3 + 1 \rangle$ is of order $2^6 - 1 = 63$ imply that t is not a generator of that group. So the polynomial $x^6 + x^3 + 1$ of $\mathbb{Z}_2[x]$ is not primitive.

Remark 4.11. If α is a primitive element of a finite field $\mathbb{F}$ with $|\mathbb{F}| = r \geq 2$, then we know that α is of order $r - 1$. In particular, α is a root of the polynomial $Q(x) = x^{r-1} - 1$ and $r - 1$ is the smallest positive integer m such that α is a root of $x^m - 1$. If $F(x)$ is a primitive polynomial of degree r over $\mathbb{Z}_p$, then $r - 1$ is the smallest positive integer n satisfying $t^n - 1 = 0$ in the field $\mathbb{Z}[x]/\langle F(x) \rangle$ and consequently $r - 1$ is the smallest positive integer n such that $F(x)$ is a divisor of $x^n - 1$.

The following theorem proves that there is enough supply of primitive polynomials of any chosen degree. The proof is omitted as it is beyond the scope of this book.

Theorem 4.10. For any prime integer p and any positive integer n, there exists a primitive polynomial of degree n over the field $\mathbb{Z}_p$.

4.5.10 *The lead function*

We start by reviewing the definition of a linear map.

Definition 4.12. A map $f : \mathbb{F}_p^r \to \mathbb{F}_p$ is called **linear** if it satisfies the two conditions:

(1) $f(\vec{u} + \vec{v}) = f(\vec{u}) + f(\vec{v})$ for all r-tuples $\vec{u}$, $\vec{v}$ in $\mathbb{F}_p^r$;
(2) $f(\alpha \vec{u}) = \alpha f(\vec{u})$ for all $\vec{u} \in \mathbb{F}_p^r$ and $\alpha \in \mathbb{F}_p$.

Example 4.20. Let $F(x)$ be an irreducible polynomial of degree r in $\mathbb{F}_p[x]$ and identify the field $\mathbb{F}_p^r$ (or $\mathbb{F}_p[x]/\langle F(x) \rangle$) as usual with the set of polynomials of degree $r - 1$ or less with coefficients in $\mathbb{F}_p$. Consider the map $\theta : \mathbb{F}_p^r \to \mathbb{F}_p$, called the **lead function**, defined as follows:

$$\theta \left(b_{r-1} t^{r-1} + \cdots + b_1 t + b_0 \right) = b_{r-1}.$$

If $\vec{u} = b_{r-1} t^{r-1} + \cdots + b_1 t + b_0$, $\vec{v} = c_{r-1} t^{r-1} + \cdots + c_1 t + c_0 \in \mathbb{F}_p^r$ and $\alpha \in \mathbb{F}_p$, then

- $\theta(\vec{u} + \vec{v}) = \theta\left((b_{r-1} + c_{r-1}) t^{r-1} + \cdots + (b_1 + c_1) t + (b_0 + c_0) \right) = b_{r-1} + c_{r-1} = \theta(\vec{u}) + \theta(\vec{v})$.
- $\theta(\alpha \vec{u}) = \theta\left(\alpha b_{r-1} t^{r-1} + \cdots + \alpha b_1 t + \alpha b_0 \right) = \alpha b_{r-1} = \alpha \theta(\vec{u})$.

This means that θ is a linear map.

Remark 4.12. A special case of great interest in our treatment of the GPS signal is the case where $p = 2$. In this case, there are 2^r polynomials of the form $b_{r-1} t^{r-1} + \cdots + b_1 t + b_0 \in \mathbb{Z}_2[t]$ with exactly half having the leading coefficient $b_{r-1} = 0$ and the other half with leading coefficient $b_{r-1} = 1$. This means that the lead function $\theta : \mathbb{F}_2^r \to \mathbb{F}_2$ takes the value 0 on exactly half of the elements of $\mathbb{F}_2^r$ and the value 1 on the other half.

4.6 Key properties of GPS signals: Correlation and maximal period

We now arrive at the last stop in our journey of understanding the mathematics behind the signal produced by a GPS satellite using a LFSR. This section provides the proof of the main result concerning the GPS signal (Theorem 4.1). We start with the notion of correlation between two "slices" of the sequence produced by a LFSR. It is the calculation of this correlation that allows the GPS receiver to accurately compute the exact time taken by the signal to reach it from the satellite.

4.6.1 *Correlation*

Definition 4.13. Given two binary finite sequences of the same length: $A = (a_i)_{i=1}^n$ and $B = (b_i)_{i=1}^n$, the **correlation** between A and B, denoted by $\nu(A, B)$, is defined as follows:

$$\nu(A, B) = \sum_{i=1}^n (-1)^{a_i} (-1)^{b_i}.$$

Let $S = \{1, 2, \ldots, n\}$, $S_1 = \{i \in S; \ a_i = b_i\}$ and $S_2 = \{i \in S; \ a_i \neq b_i\}$. Then

$$\sum_{i=1}^n (-1)^{a_i} (-1)^{b_i} = \sum_{i \in S_1} (-1)^{a_i} (-1)^{b_i} + \sum_{i \in S_2} (-1)^{a_i} (-1)^{b_i}.$$

Note that:

- if $a_i = b_i$, then $(-1)^{a_i}(-1)^{b_i} = (-1)^{2a_i} = 1$, so $\sum_{i \in S_1} (-1)^{a_i}(-1)^{b_i} = 1 + 1 + \cdots + 1$ as many times as the number of elements in S_1.
- if $a_i \neq b_i$, then $(-1)^{a_i}(-1)^{b_i} = -1$ since one of a_i, b_i is 0 and the other is 1 in this case. We conclude that $\sum_{i \in S_2} (-1)^{a_i}(-1)^{b_i} = -1 - 1 - \cdots - 1$ as many times as the number of elements in S_2.

Thus, the correlation between A and B is equal to the number of elements in S_1 minus that of S_2. This proves the following proposition.

Proposition 4.3. The correlation between two binary finite sequences $A = (a_i)_{i=1}^n$ and $B = (b_i)_{i=1}^n$ of the same length is equal to the number of indices i where $a_i = b_i$ minus the number of indices i where $a_i \neq b_i$.

The above proposition suggests that the correlation serves as a measure how "similar" the sequences are. The closer the correlation $\nu(A, B)$ to zero, more corresponding terms in the sequences are different. So sequences with small correlation are poorly correlated and sequences with large correlation are strongly correlated.

Example 4.21. Consider the following two finite sequences:

$$101011100101110$$
$$111001011100101.$$

Every time the numbers agree (in black), add 1 and every time the numbers disagree (in gray), subtract 1. The resulting correlation is then -1.

4.6.2 *The LFSR sequence revisited*

In this section, we revisit the sequence produced by a LFSR of degree r (with r cells) and try to analyse it in a bit more depth. We start by fixing a primitive polynomial of degree r over $\mathbb{Z}_2$:

$$P(x) = x^r + c_{r-1}x^{r-1} + \cdots + c_1x + c_0.$$

We know that such a polynomial exists by Theorem 4.10 above. For the coefficients of the LFSR, choose the vector $c = (c_{r-1}, \ldots, c_1, c_0)$ with components equal to the coefficients of $P(x)$. The choice of the initial window can be any non-zero vector $(a_{r-1}, \ldots, a_1, a_0)$ but the one we choose in the following is very suitable in proving interesting facts about the sequence. We use the lead function $\theta : \mathbb{F}_p^r \to \mathbb{F}_p$ defined in Example 4.20 above as follows:

(1) Choose a non-zero polynomial $\epsilon(t)$ in $\mathbb{Z}_2[x]/\langle P(x)\rangle$:

$$\epsilon(t) = \epsilon_{r-1}t^{r-1} + \cdots + \epsilon_1 t + \epsilon_0, \quad \epsilon_i \in \mathbb{Z}_2 \text{ for all } i = r-1, \ldots, 0.$$

(2) Define $a_0 = \theta(\epsilon(t)) = \epsilon_{r-1}$.
(3) Next, we compute $t\epsilon(t)$ as an element of $\mathbb{Z}_2[x]/\langle P(x)\rangle$. Remember that the equation $P(t) = 0$ together with the fact that $-c_i = c_i$ in the field $\mathbb{Z}_2$ translates to $t^r = c_{r-1}t^{r-1} + \cdots + c_1 t + c_0$.

$$\begin{aligned}
t\epsilon(t) &= t\left(\epsilon_{r-1}t^{r-1} + \cdots + \epsilon_1 t + \epsilon_0\right) \\
&= \epsilon_{r-1}t^r + \epsilon_{r-2}t^{r-1} \cdots + \epsilon_1 t^2 + \epsilon_0 t \\
&= \epsilon_{r-1}\left(c_{r-1}t^{r-1} + \cdots + c_1 t + c_0\right) + \epsilon_{r-2}t^{r-1} \cdots + \epsilon_1 t^2 + \epsilon_0 t \\
&= \left(\epsilon_{r-1}c_{r-1} + \epsilon_{r-2}\right)t^{r-1} + \cdots + \left(\epsilon_{r-1}c_1 + \epsilon_0\right)t + \epsilon_{r-1}c_0.
\end{aligned}$$

(4) Define $a_1 = \theta(t\epsilon(t)) = \epsilon_{r-1}c_{r-1} + \epsilon_{r-2}$.

(5) To define a_2, we compute first $t^2\epsilon(t)$ as an element of $\mathbb{Z}_2[x]/\langle P(x)\rangle$ (always using the identity $t^r = c_{r-1}t^{r-1} + \cdots + c_1 t + c_0$) and then we define a_2 as the lead of the resulting polynomial. We find $a_2 = \theta(t^2\epsilon(t)) = \epsilon_{r-1}\left(c_{r-1}^2 + c_{r-2}\right) + \epsilon_{r-2}c_{r-1}$.

(6) In general, define $a_i = \theta(t^i\epsilon(t))$ for all $i \in \{0, 1, \ldots, r-1\}$.

(7) Take $(a_{r-1}, \ldots, a_1, a_0)$ to be the initial window of the LFSR.

But what is the big deal? Why do we need $P(x)$ to be primitive and why this complicated way of choosing the initial window? Be patient, you have gone a long way so far and the answers are just a few paragraphs away.

Note that:

$$\theta\left(t^r\epsilon(t)\right) = \theta\left(c_{r-1}t^{r-1}\epsilon(t) + \cdots + c_1 t\epsilon(t) + c_0\epsilon(t)\right) \tag{4.8}$$

$$= c_{r-1}\theta\left(t^{r-1}\epsilon(t)\right) + \cdots + c_1\theta\left(t\epsilon(t)\right) + c_0\theta\left(\epsilon(t)\right) \tag{4.9}$$

$$= c_{r-1}a_{r-1} + \cdots + c_1 a_1 + c_0 a_0, \tag{4.10}$$

where (4.8) follows from $t^r = c_{r-1}t^{r-1} + + \cdots + c_1 t + c_0$, (4.9) follows from the linearity of the lead map θ and (4.10) follows from our definition of the initial conditions $a_0, \ldots, a_{r-1}$. Look closely at the last expression. Is that not how the LFSR computes its next term a_r? We conclude that $\theta(t^r\epsilon(t)) = a_r$. In fact, it is not hard to show that any term in the sequence produced by a LFSR can be obtained this way. More specifically,

$$a_k = \theta(t^k\epsilon(t)), \quad k = 0, 1, 2, \ldots. \tag{4.11}$$

4.6.3 *Proof of Theorem 4.1*

We are now ready to prove Theorem 4.1.

Proof of Theorem 4.1. With the above choice of the coefficients (as coefficients of a primitive polynomial) and the initial coefficients, we show that a sequence produced by a LFSR with r registers has a period equal to $N = 2^r - 1$. We already know (see Remark 4.1) that the sequence is periodic with (minimal) period $T \leq 2^r$. Since $P(x)$ is chosen to be a primitive polynomial, t is a generator of the multiplicative group of the field $\mathbb{Z}_2[x]/\langle P(x)\rangle$ and so it has an order of $N = 2^r - 1$ as an element of the group. This means that N is the smallest positive integer satisfying $t^N = 1$. Moreover, for any $n \in \mathbb{N}$, we have

$$a_{n+N} = \theta\left(t^{n+N}\epsilon(t)\right) = \theta\left(\underbrace{t^N}_{=1} t^n\epsilon(t)\right) = \theta\left(t^n\epsilon(t)\right) = a_n.$$

This shows in particular that N is a period of the sequence and by the minimality of the period T, we have that

$$T \leq N. \tag{4.12}$$

For any $k \in \mathbb{N}$, if we apply θ to the relation $a_{k+T} = a_k$ we get $\theta\left(t^{k+T}\epsilon(t)\right) = \theta\left(t^k\epsilon(t)\right)$ or equivalently

$$\theta\left(t^k\epsilon(t)(t^T - 1)\right) = 0 \tag{4.13}$$

by the linearity of θ. Assume $(t^T - 1) \neq 0$, then $\epsilon(t)(t^T - 1) \neq 0$ as the product of two non-zero elements of the field $\mathbb{Z}_2[x]/\langle P(x)\rangle$. The polynomial $P(x)$ was chosen to be minimal for a reason: any non-zero element of $\mathbb{Z}_2[x]/\langle P(x)\rangle$ is a power of t, in particular $\epsilon(t)(t^T - 1) = t^n$ for some $n \in \{0, 1, 2, \ldots, N - 1\}$ and therefore $t^k\epsilon(t)(t^T - 1) = t^{k+n}$. The elements $t^k\epsilon(t)(t^T - 1)$ are then just permutations of the elements of multiplicative group $\mathbb{F}_{2^r}^* = \{1, t, t^2, \ldots, t^{N-1}\}$. Equation (4.13) implies that the lead function θ takes the value zero everywhere on $\mathbb{F}_{2^r}^*$ which is not true (see Remark 4.12 above). Therefore $t^T - 1 = 0$ or equivalently $t^T = 1$. Since N is the order of t as element of the multiplicative group of the field $\mathbb{Z}_2[x]/\langle P(x)\rangle$, we get that $N \leq T$. Using inequality (4.12), we get that $T = N$. We conclude that the (minimal) period of the sequence a_n is indeed $N = 2^r - 1$. This finishes the proof of the theorem.

Example 4.22. It can be shown that the polynomial $P(x) = x^4 + x^3 + 1$ is primitive over $\mathbb{Z}_2$. We look at the binary sequence produced by a LFSR of degree 4 using the polynomial $P(x) = x^4 + x^3 + 1$. The coefficient vector of the LFSR in this case is $(c_3, c_2, c_1, c_0) = (1, 0, 0, 1)$. For the initial values, let us choose the non-zero polynomial $\epsilon(t) = t^2 + t$. We compute $t\epsilon(t)$, $t^2\epsilon(t)$ and $t^3\epsilon(t)$ as elements of the field $\mathbb{Z}_2[x]/\langle P(x)\rangle$. Remember first that in the field $\mathbb{Z}_2[x]/\langle x^4 + x^3 + 1\rangle$, we have the identity $t^4 + t^3 + 1 = 0$ or $t^4 = -t^3 - 1 = t^3 + 1$.

$$t\epsilon(t) = t^3 + t^2,\ t^2\epsilon(t) = t^4 + t^3 = (t^3 + 1) + t^3 = 2t^3 + 1 = 1,\ t^3\epsilon(t) = t.$$

So $a_0 = \theta(\epsilon(t)) = 0$, $a_1 = \theta(t\epsilon(t)) = 1$, $a_2 = \theta(t^2\epsilon(t)) = 0$ and $a_3 = \theta(t^3\epsilon(t)) = 0$. The initial state vector is then $(a_3, a_2, a_1, a_0) = (0, 0, 1, 0)$.

The following tables give the first 24 windows of the sequence:

Clock Pulse #	Window	Clock Pulse #	Window	Clock Pulse #	Window
0	0010	8	0101	16	0001
1	0001	9	1010	17	1000
2	1000	10	1101	18	1100
3	1100	11	0110	19	1110
4	1110	12	0011	20	1111
5	1111	13	1001	21	0111
6	0111	14	0100	22	1011
7	1011	15	0010	23	0101

The sequence produced is ... 111000100110101111100010 which is periodic of period $2^4 - 1 = 15$. When the sequence is read from right to left, one can see that the slice 011010111100010 is repeated every 15 bits.

We can actually say more about the sequence produced by a LFSR as constructed above.

Theorem 4.11. Consider the binary sequence produced by a LFSR of degree r constructed as above. Let $W_1 = (a_n, a_{n+1}, \ldots, a_{n+N-1})$ and $W_2 = (a_m, a_{m+1}, \ldots, a_{m+N-1})$ be two slices (or subsequences) of the sequence with $m > n$ and length $N = 2^r - 1$ (the minimal period of the sequence) each. Then the correlation ν between W_1 and W_2 is given by:

$$\nu = \begin{cases} -1 \text{ if } m - n \text{ is not a multiple of } N \\ N \text{ if } m - n \text{ is a multiple of } N. \end{cases}$$

Proof. We use the definition of the correlation,

$$\nu(W_1, W_2) = \sum_{k=0}^{N-1} (-1)^{a_{n+k}} (-1)^{a_{m+k}}$$

$$= \sum_{k=0}^{N-1} (-1)^{\theta(t^{n+k}\epsilon(t))} (-1)^{\theta(t^{m+k}\epsilon(t))} \quad \text{(By relation (4.11) above)}$$

$$= \sum_{k=0}^{N-1} (-1)^{[\theta(t^{n+k}\epsilon(t)) + \theta(t^{m+k}\epsilon(t))]}.$$

By the linearity of the lead function, we get

$$\nu(W_1, W_2) = \sum_{k=0}^{N-1} (-1)^{\theta(t^{n+k}\epsilon(t) + t^{m+k}\epsilon(t))} = \sum_{k=0}^{N-1} (-1)^{\theta(t^{n+k}\epsilon(t)(1+t^{m-n}))}.$$

If $m - n$ is a multiple of N, then $m - n = \rho N$ for some integer ρ and $t^{m-n} = (t^N)^\rho = 1$ since $t^N = 1$ (remember that t is the generator of a cyclic group of order N). So, $1 + t^{m-n} = 2 = 0$ and $(-1)^{\theta\left(t^{n+k}\epsilon(t)\left(1+t^{m-n}\right)\right)} = 1$ for all $k = 0, \ldots, N-1$. This implies that the correlation in this case is $\nu = \underbrace{1 + 1 + \cdots + 1}_{N} = N$. Assume next that $m - n$ is not a multiple of N, then the polynomial $1 + t^{m-n}$ is non-zero and therefore $\epsilon(t)\left(1 + t^{m-n}\right)$ is also non-zero as the product of two non-zero elements of the field $\mathbb{Z}_2[x]/\langle P(x)\rangle$. As in the proof of Theorem 4.1, the fact that $P(x)$ is chosen to be primitive comes in very handy now: $\epsilon(t)\left(1 + t^{m-n}\right) \neq 0$ implies that

$$\epsilon(t)\left(1 + t^{m-n}\right) = t^j \text{ for some } j \in \{0, 1, 2, \ldots, N-1\}.$$

As k takes all values in the set $\{0, 1, \ldots, N-1\}$, the elements $t^{n+k}\epsilon(t)\left(1 + t^{m-n}\right) = t^{j+n+k}$ are just permutations of the elements of $\mathbb{F}_{2^r}^* = \{1, t, t^2, \ldots, t^{N-1}\}$. As seen above, the lead function θ takes the value 0 on exactly half of the elements of the set $\mathbb{F}_{2^r}$ and the value 1 on the other half. This implies in particular that $\sum_{\alpha_i \in \mathbb{F}_{2^r}} (-1)^{\theta(\alpha_i)} = 0$. Now, since $(-1)^{\theta(0)} = (-1)^0 = 1$, the last sum in the above expression of $\nu(W_1, W_2)$ can be written as

$$\sum_{k=0}^{N-1} (-1)^{\theta\left(t^{n+k}\epsilon(t)\left(1+t^{m-n}\right)\right)} = \sum_{\alpha_i \in \mathbb{F}_{2^r}^*} (-1)^{\theta(\alpha_i)}$$

$$= \underbrace{\sum_{\alpha_i \in \mathbb{F}_{2^r}} (-1)^{\theta(\alpha_i)}}_{0} - (-1)^{\theta(0)} = -1.$$

This proves that the correlation between the two finite sequences is -1 in this case. $\qquad\square$

This is indeed an amazing fact: take any two finite slices of the same length $2^r - 1$ (length of a period) in a sequence produced by a LFSR of degree r, then you are sure that the number of terms which disagree is always one more than the number of terms which agree (provided, as in the theorem, that $m - n$ is not a multiple of $N = 2^r - 1$). This may sound weird, but having poorly correlated sequences of maximal length $2^r - 1$ is important for the GPS receiver since it makes the task of identifying satellites much easier.

4.6.4   *More about the signal*

Sequences produced by LFSR on board of GPS satellites are modulated by wave carriers and transformed into cycles of electrical "chips" or pulses that we usually represent by sequences of 0's and 1's for simplicity. These are just representations of low and high voltages.

Upon reception of a signal, the receiver tries immediately to match it with one of the local replicas of the codes stored in its memory. As explained earlier, the received code is not synchronized with any of the locally generated codes because of the runtime of the signal from the satellite and the fact that the satellites are in constant movement. Once the receiver succeeds to match the received code with one of the replicas, it can identify the satellite from which the signal was emitted and starts to collect the information needed to determine the travel time of the signal. To achieve synchronization, the receiver shifts its locally generated signal by one chip at a time and compares it with the captured signal by calculating the correlation between two cycles of the codes. This process is repeated until a maximal correlation is attained (and hence perfect synchronization between the two signals).

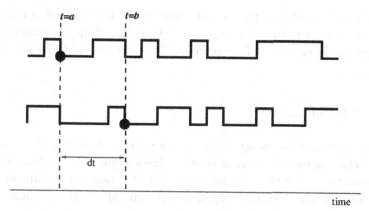

In the above diagram, the first signal represents the replica of the satellite code generated by the receiver with $t = a$ being the time of departure of a particular code cycle from the satellite. The second signal represents the signal arriving at the receiver with $t = b$ being the time of arrival of the cycle to the receiver. Signals emitted by various GPS satellites are in perfect synchronization and the departure time from the satellites of the start of each cycle is known by the receiver. The runtime of the signal is marked by dt. In a perfect scenario, the distance between the satellite and the receiver is $c \cdot dt$ with $c = 299792.458$ km/s being the speed of light in a vacuum. Unfortunately, many factors play a role in making the calculation of the distance a bit more complicated than that. For instance, remember that a major source of error comes from the fact that clocks on board of the satellites and inside the receiver read different times at any given moment. But that can be fixed by looking at the signal of a fourth satellite as explained in Section 4.4.5. To compute the time offset uncertainty, the receiver records the number n of electrical chips needed to be shifted in order to achieve synchronization (when the correlation is at a maximum) between the two signals. On board of a GPS satellite, the LFSR used to produce the PNR code is of degree 10 (has $r = 10$ registers), producing a sequence of period $2^{10} - 1 = 1023$ bits by the above discussion. Practically, this means that each cycle of the satellite signal is formed by 1023 electrical chips. The code is generated at a rate of 1.023 megabits/sec which means that a cycle repeats every millisecond (or 0.001 second). At the speed of light, 0.001 second corresponds to a distance of 299.792458 km. Dividing this distance with 1023 (the period of the sequence) gives an uncertainty of about 300 m per chip.

In reality, analyzing the satellite code at the receiver end is more sophisticated than the above description. Many algorithms are implemented to increase the efficiency of the receiver. These are beyond the scope of this book.

4.7 A bit of history

The idea of locating one's position on the surface of the planet goes back deep in human history. Some ancient civilizations were able to develop navigational tools (like the Astrolabe) to locate the position of ships in high seas. But let us not go that deep in history, after all the chapter deals with a very recent piece of technology.

In what follows, we give a brief history of the development of the Global Positioning System.

- The story started in 1957 when the Soviet Union launched the satellite *Sputnik*. Just days after, two American scientists were able to track its orbit simply by recording changes in the satellite radio frequency.
- In the 1960's, the American navy designed a navigation system for its

fleet consisting of 10 satellites. At that time, the signal reception was very slow and scientists worked hard to improve it.

- In the early 1970's, Ivan Getting and Bradford Parkinson led a US defense department project to provide continuous navigation information, leading to the development of the GPS (formally known as NAVSTAR GPS) in 1973.

- The launching of the first GPS satellite by the US military was accomplished in 1978.

- In 1980, the military activated atomic clocks on board of GPS satellites.

- The year 1983 was a turning point in the development of the GPS system. It was the year where the system started to take the form that we use today. This came after the tragedy of flight 007 of the Korean Airline that killed 269 people on board. The tragedy prompted US president Ronald Reagan to declassify the GPS project and cleared the way to allow the civilian use of the system.

- After many setbacks, full operational capability with 24 GPS satellites was announced in 1995.

- In May 2000, the Selective Availability program was discontinued following an executive order issued by US president Bill Clinton to make the GPS more responsive to civilian and commercial users worldwide. This created a boom in the GPS devices production industry.

- In 2005, the GPS constellation consisted of 32 satellites, out of which 24 are operational and 8 are ready to take over in case of failure.

- In 2009, an alarming report by the US Accountability Office warned that some GPS satellites could fail as early as 2010.

- The year 2010 marked an important step in the modernization of the GPS system with the announcement of the US government of a new contract to develop the GPS Next Generation Operational Control System.

4.8   References

Elliott Kaplan and Christopher Hegarty (editors) (2005). *Understanding GPS: Principles and Applications*, Second Edition. (Artech House).

Christiane Rousseau, Yvan Saint-Aubin (2008) *Mathematics and Technology.* (Springer).

Chapter 5

Image processing and face recognition

5.1 Introduction

In this chapter, we discuss the processing of images. Chances are you have used at some point an image editing software like Photoshop or Gimp to transform a photo. We discuss some of the mathematics underlying the manipulation of images. Our approach will be to combine many images to obtain the average image. Consider the following three digital images. The image on the left is a picture taken at the Tremblant ski hill in 2015. The image in the center is a picture taken in Vancouver in 2012 and the image on the right is a picture taken in Morocco in 2015. We will combine the faces from these images to obtain an "average" face.

Many police organizations have a database of facial images, which is used to identify individuals who have been involved in a crime. The identification needs to be automated, since typically, the database will contain a large number of images. We discuss a technique that compares an image with images in a database for facial recognition.

5.1.1 *Before you go further*

Mathematical skills required to have a good understanding of this chapter include basic knowledge of descriptive statistics, basic linear algebra concepts like the dot product and orthogonal projection. Also, some good knowledge of working with and manipulating matrices is necessary.

5.2 Raster image

There are many forms of digital images. Our goal here is to discuss the manipulation of raster images. We can think of a raster image as an array of pixels, where the pixel represents a unit square. For example, the above image in Morocco has a height of 960 pixels and width of 720 pixels. The image in Vancouver has a height of 720 pixels and a width of 960 pixels. Lastly, the image in Tremblant is 960×540 squares pixels.

As seen in Chapter 3, a gray level is assigned to each pixel. The levels are usually from 0 to 255, where 0 is black and 255 is white. Alternatively, we can think of the levels as percentages, where 0% is black and 100% is white.

Each black and white digital image can be represented by a matrix of gray levels. For example, the image in Figure 5.1 is represented by the matrix $C = \begin{bmatrix} 10 & 35 & 100 \\ 125 & 175 & 200 \end{bmatrix}$ where, for instance, the upper left pixel is assigned the gray level 10 and the lower right pixel is assigned the level 200.

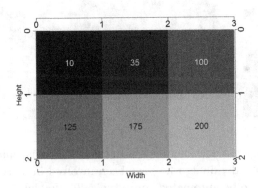

Fig. 5.1 A raster image.

Transformations of the raster images correspond to transformations of the corresponding matrix. The notation $C[i,j]$ is used to indicate the color (or gray level) in the ith row and the jth column of the image.

There are many systems to represent colored images. A widely used one is the RED-GREEN-BLUE (RGB) color space. In the RGB color space, each pixel is represented by a triplet (R,G,B) of integers in $[0,255]$ with $(R,G,B) = (0,0,0)$ corresponding to black and $(R,G,B) = (255,255,255)$ corresponding to white. A raster image in the RGB color space will have three matrices, one for each color. In order to transform a color image, one needs to transform each of the three matrices. For simplicity, we only look at the transformation of grayscale images in this chapter.

5.3 Invertible linear transformations

We will interpret a point in our image as an element (x,y) in $\mathbb{R}^2$. We usually consider the origin as the upper left corner of the image (recall the image is rectangular). Increasing the first component (on the x-axis) by one unit means moving to the right by one unit. Increasing the second component (on the y-axis) by one unit means moving down by one unit. A unit square is just one pixel. The reason to have the upper left corner as the origin is the need to have an easy correspondence between the points on the image and the matrix of the image. The color of the point (x,y) in the image is $C[\lceil y \rceil, \lceil x \rceil]$, where $\lceil x \rceil$ is the smallest integer greater than or equal to x (it is rounding up to the nearest integer).

For example, the point $(0.5, 0.2)$ in Figure 5.1 is within the first pixel of the image. Since the gray level of the first pixel is 10, we have $C[\lceil 0.2 \rceil, \lceil 0.5 \rceil] = C[1,1] = 10$.

In this chapter we discuss rotations and uniform scaling, since these are the transformations used to align faces. Recall that the origin $(0,0)$ is the point in the upper left corner of the rectangular image. However, we will want to apply the transformation (e.g. a rotation) about a point in the facial region, say the point (x_0, y_0), that we call the **centroid**. We represent a point (x,y) on the image with a coordinate system with respect to the centroid, by using the coordinates $[v_x, v_y] = [x - x_0, y - y_0]$. For example, if the centroid is the point $(2,2)$, then the point $(2,1)$ can also be represented as $[2 - 2, 1 - 2] = [0, -1]$. The color of $[v_x, v_y]$ is $C[\lceil v_y + y_0 \rceil, \lceil v_x + x_0 \rceil]$, since it corresponds to the point $(v_x + x_0, v_y + y_0) = (x,y)$.

Consider the image delimited by the rectangle $ABCD$ in Figure 5.2, which is 3 pixels wide and 4 pixels high. The grayscale matrix of this image is

$$C = \begin{bmatrix} 25 & 45 & 55 \\ 45 & 75 & 125 \\ 55 & 175 & 200 \\ 99 & 190 & 180 \end{bmatrix}. \tag{5.1}$$

Using the point $(2, 2)$ as the centroid. The point $(x, y) = (0.5, 1.5)$ has also the following coordinates $[v_x, v_y] = [0.5 - 2, 1.5 - 2] = [-1.5, -0.5]$ with respect to the centroid. The color of this point is $C[\ulcorner 1.5 \urcorner, \ulcorner 0.5 \urcorner] = C[2, 1] = 45$. The computation of the color when referring to the coordinates of the point with respect to the centroid is $C[\ulcorner -1.5 + 2 \urcorner, \ulcorner -0.5 + 2 \urcorner] = C[2, 1] = 45$.

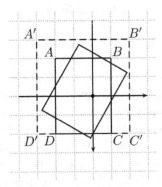

Fig. 5.2   Rotation of an image by $\pi/6$ radians.

There are many transformations in Photoshop or Gimp that are invertible linear maps: rotation about a point, reflection about a line, a shear, a uniform scaling, and a non-uniform scaling. Recall that a linear transformation from $\mathbb{R}^2$ to $\mathbb{R}^2$ is a function $T : \mathbb{R}^2 \to \mathbb{R}^2$ such that $T(\vec{v}) = A_T \vec{v}$, for some 2×2 matrix A_T called the matrix of the linear transformation. In other words, if $A_T = \begin{bmatrix} a & b \\ c & d \end{bmatrix}$, then the image $\vec{v}' = [v'_x, v'_y]$ of $\vec{v} = [v_x, v_y]$ (i.e. location in the new image) is given by

$$\begin{pmatrix} v'_x \\ v'_y \end{pmatrix} = \begin{bmatrix} a & b \\ c & d \end{bmatrix} \begin{pmatrix} v_x \\ v_y \end{pmatrix} = \begin{pmatrix} a\,v_x + b\,v_y \\ c\,v_x + d\,v_y \end{pmatrix}.$$

The matrix of the linear transformation that rotates the image by an angle of θ radians (about the point (x_0, y_0), clockwise) is known to be

$$R_\theta = \begin{bmatrix} \cos(\theta) & -\sin(\theta) \\ \sin(\theta) & \cos(\theta) \end{bmatrix}.$$

Its inverse R_θ^{-1} is the rotation matrix $R_{-\theta}$. In other words, to invert the rotation, we apply a rotation with the same angle in the opposite direction.

To uniformly scale the image by a factor of r (where $r > 0$), we can use the following matrix

$$S_r = \begin{bmatrix} r & 0 \\ 0 & r \end{bmatrix} = r \begin{bmatrix} 1 & 0 \\ 0 & 1 \end{bmatrix}.$$

So $S_r = r\,I_2$, where I_2 is the identity matrix of size 2×2. The inverse of the uniform scaling matrix S_r is

$$S_r^{-1} = S_{1/r} = \begin{bmatrix} 1/r & 0 \\ 0 & 1/r \end{bmatrix} = \frac{1}{r} \begin{bmatrix} 1 & 0 \\ 0 & 1 \end{bmatrix} = \frac{1}{r} I_2.$$

When applying both a uniform scaling by a factor r and a rotation of angle θ, the order of the transformations is not important. The matrix for the transformation that starts with a rotation followed by a uniform scaling is $A_T = S_r\,R_\theta$. This matrix is equal to

$$S_r\,R_\theta = r\,I_2\,R_\theta = r\,R_\theta = r\,R_\theta\,I_2 = R_\theta\,(r\,I_2) = R_\theta\,S_r.$$

The last matrix in this equality is the matrix for the transformation that is a uniform scaling followed by the rotation. So regardless of the order of these two transformations the result is the same. In fact the matrix of the linear transformation that corresponds to rotation of angle θ clockwise and a uniform scaling of a factor r is

$$r\,R_\theta = \begin{bmatrix} r\cos(\theta) & -r\sin(\theta) \\ r\sin(\theta) & r\cos(\theta) \end{bmatrix} = \begin{bmatrix} a & -b \\ b & a \end{bmatrix},$$

where $a = r\cos(\theta)$ and $b = r\sin(\theta)$. Theorem 5.1 states that any matrix of the same form as the matrix on the right-hand-side of the above equality can be interpreted as a matrix of the composition of a rotation and a uniform scaling.

Before stating the theorem, we remind the reader that an ordered pair $(x, y) \in \mathbb{R}^2$ (which is not $(0, 0)$) has a unique polar coordinate representation (r, θ), where $r > 0$ and $\theta \in (-\pi, \pi]$. There is a one-to-one correspondence between the cartesian coordinates (x, y) and the polar coordinates (r, θ). To go from the polar form to the cartesian form, we use the relations

$x = r\cos(\theta)$ and $y = r\sin(\theta)$. To go from the cartesian form to the polar form, we use $r = \sqrt{x^2 + y^2}$ and $\theta = \text{atan}_2(y, x)$, where

$$\text{atan}_2(y, x) = \begin{cases} \arctan(y/x), & \text{if } x > 0 \\ \arctan(y/x) + \pi, & \text{if } y \geq 0, x < 0 \\ \arctan(y/x) - \pi, & \text{if } y < 0, x < 0 \\ \pi/2, & \text{if } y > 0, x = 0 \\ -\pi/2, & \text{if } y < 0, x = 0 \\ \text{undefined}, & x = 0, y = 0 \end{cases}$$

The atan_2 function is sometimes called the "Four-Quadrant Inverse Tangent". Its domain is $\mathbb{R}^2 \setminus \{(0,0)\}$ and its image is $(-\pi, \pi]$. It is implemented in most modern computer languages and it is a convenient way to pass from cartesian form to polar form.

Theorem 5.1. Let T be linear map from $\mathbb{R}^2$ to $\mathbb{R}^2$ with the matrix

$$A_T = \begin{bmatrix} a & -b \\ b & a \end{bmatrix},$$

where $a, b \in \mathbb{R}$ and $(a, b) \neq (0, 0)$. Then T can be interpreted as a rotation of angle θ and a uniform scaling by a factor r, where

$$r = \sqrt{a^2 + b^2}, \quad \theta = \text{atan}_2(b, a).$$

Proof. Since (r, θ) is the polar coordinate for (a, b), then $a = r\cos(\theta)$ and $b = r\sin(\theta)$. Thus, we can write $A_T = r\,R_\theta$, which is a clockwise rotation of angle θ and a uniform scaling of factor r. $\qquad\square$

When applying a linear transformation (e.g. a rotation), we might end up mapping some points outside of the region of the given image. As an example, consider the rotation of the image in Figure 5.2. The boundary of the region of the image corresponds to the rectangle $ABCD$. As we rotate the image by an angle $\pi/6$ radians clockwise about the point $(2, 2)$, some points are mapped outside of the rectangle $ABCD$. In order not to lose information, we increase the number of pixels. The new image, that corresponds to the rectangle $A'B'C'D'$, will be of size 5×5 pixels, while the original image was of size 3×4.

We first discuss the *math* involved in determining the new rectangle $A'B'C'D'$ and identifying the corresponding centroid (x_0', y_0') for our new image. The original rectangle $ABCD$ is a convex set and the points A, B, C and D are the only extremal points in this set. (Refer to Section 5.8 for a discussion concerning convex sets.) If we are using an invertible linear transformation to transform the image, then the region of the new image will also be a convex set and the images of A, B, C and D, say $T(A) = [a_x, a_y]$, $T(B) = [b_x, b_y]$, $T(C) = [c_x, c_y]$, $T(D) = [d_x, d_y]$, respectively, are going to be the only extremal points in this set. So to find the boundary of the new image, we need the extreme abscissa:

$$x'_{min} = \min\{a_x, b_x, c_x, d_x\} \quad \text{and} \quad x'_{max} = \max\{a_x, b_x, c_x, d_x\}.$$

We will also need the extreme ordinates:

$$y'_{min} = \min\{a_y, b_y, c_y, d_y\} \quad \text{and} \quad y'_{max} = \max\{a_y, b_y, c_y, d_y\}.$$

We will round these extreme values to the nearest integer to work in pixels. (We might lose a bit of information by rounding, but it should be negligible, i.e. not visible.) In Figure 5.2, $x'_{min} = -3$, $x'_{max} = 2$, $y'_{min} = -3$, $y'_{max} = 2$, measured from the centroid $(2, 2)$. Thus, the vertices of the new rectangular image are $A' = [-3, -3]$, $B' = [2, -3]$, $C' = [2, 2]$, $D' = [-3, 2]$, respectively. So, the new rectangle should be $x'_{max} - x'_{min} = 5$ pixels wide and $y'_{max} - y'_{min} = 5$ pixels high. Since A' is the point in the upper left corner of the new image, it should correspond to the point $(0, 0)$. But its coordinates with respect to the centroid $(2, 2)$ are given by the point $A' = [-3, -3]$, which means that it is 3 unit above and 3 units left of the centroid. By making the centroid the point $(3, 3)$, then A' will correctly be the point $(0, 0)$. So the centroid in the new image is the point $(3, 3)$.

5.4 Gray level for the new image

Now that we have the appropriate rectangular region for the transformed image, we must color it. In other words, we must determine the gray level for each pixel in the image. Suppose that the new image has a width of w' pixels and a height of h' pixels (in Figure 5.2, we have $w' = 5$ and $h' = 5$). It is important to keep in mind the center of the image. In case of Figure 5.2, it is $(x'_0, y'_0) = (3, 3)$. We need to find $C'[i, j]$ for $i = 1, 2, \ldots, h'$ and $j = 1, 2, \ldots, l'$, where C' is the matrix of the transformed image.

To find the color, we must think in terms of the inverse transformation. $C'[i, j]$ is the color of a pixel, which is a unit square. We will use the point $(j - 1/2, i - 1/2)$ (which is the center of the corresponding pixel) as a representative of the pixel. As an example, let us consider the gray level $C'[3, 4]$ in the new image. This is the color of the pixel which has $(x', y') = (3.5, 2.5)$ as its representative. Expressing the point in terms of coordinates with respect to the centroid $(3,3)$ gives us $[3.5 - 3, 2.5 - 3] = [0.5, -0.5]$. Let us back-transform this vector.

The transformation from the original image to the new image is a rotation of $\pi/6$ radians about the centroid. The inverse transformation is a rotation of $-\pi/6$ radians about the center. The point with coordinates $\vec{v}' = [0.5, -0.5]$ with respect to the centroid $(3,3)$ has the pre-image

$$\vec{v} = R_{-\pi/6}\vec{v}' = \begin{bmatrix} \cos(-\pi/6) & -\sin(-\pi/6) \\ \sin(-\pi/6) & \cos(-\pi/6) \end{bmatrix} \begin{bmatrix} 0.5 \\ -0.5 \end{bmatrix} \approx \begin{bmatrix} 0.183 \\ -0.683 \end{bmatrix}.$$

Recall that the centroid in the original image is the point $(2,2)$. So the representative of our pixel is located at the point $(2.183, 1.317)$. The unit square centered at $(2.183, 1.317)$, which is highlighted in Figure 5.3, is overlapping four pixels in the original image (i.e. the rectangular region $ABCD$). Naturally, we should use a combination of the four colors: $C[1,2] = 45$, $C[1,3] = 55$, $C[2,2] = 75$, and $C[2,3] = 125$. The largest overlap is with the pixel in the second row and the third column of the rectangular region $ABCD$. Somehow, the color should be closest to $C[2,3] = 125$.

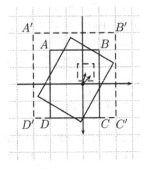

Fig. 5.3   The pixel centered at $[0.5, -0.5]$ is back transformed.

5.5   Bilinear interpolation

In this section, we discuss the assignment of a color $C[i*, j*]$ to our pixel in the case where $i*$ and $j*$ are not integers. This corresponds to the case where the back transformed pixel is overlapping pixels in the original image. To this end, we use a technique called bilinear interpolation.

Consider the color $C[i*, j*] = C[i + p, j + q]$, where i, j are integers and $0 \le p < 1$, $0 \le q < 1$. Its bilinear interpolation is

$$C[i*, j*] = (1 - p)(1 - q) C[i, j] + (1 - p) q C[i, j + 1]$$
$$+ p(1 - q) C[i + 1, j] + pq C[i + 1, j + 1].$$

We should note that it is a linear function in (p, q) with a bilinear term, that is pq. In fact, it is a weighted average of the four colors, where the weights are relative to the overlapping area with our back transformed pixel. The more the corresponding pixel overlaps our back transformed pixel, the larger its weight.

Note that the lower right point in a pixel is equivalent to referring to the pixel in terms of its column and its row. For example, in the rectangular region $ABCD$ from Figure 5.3, the lower right point of the pixel in the 3rd column and the second row is the point $(3, 2)$. We want the color of the pixel whose representative is $(2.183, 1.317)$. Its lower right corner is $(2.183 + 0.5, 1.317 + 0.5) = (2.683, 1.817)$. This pixel is highlighted in Figure 5.3. It is overlapping four pixels in the original image (i.e. the rectangular region $ABCD$). We will compute the gray level $C[1.817, 2.683]$ by using the following four gray levels:

$$C[1, 2] = 45, C[1, 3] = 55, C[2, 2] = 75, \text{ and } C[2, 3] = 125.$$

We could start by assuming that the row was an integer, say $i = 1$ or $i = 2$. In both of those cases, consider a linear approximation in j:

$$C[1, 2.683] \approx (1 - 0.683) C[1, 2] + 0.683 C[1, 3] = 51.83,$$

and

$$C[2, 2.683] \approx (1 - 0.683) C[2, 2] + 0.683 C[2, 3] = 109.15.$$

The last step is to do a linear approximation in i:

$$C[1.817, 2.683] \approx (1 - 0.817) C[1, 2.683] + .817 C[2, 2.683] \approx 77.$$

These steps give us the bilinear interpolation $C[1.317, 2.183]$, which is a weighted average of the four colors:

$$77 \approx (5.8\%) C[1, 2] + (12.5\%) C[1, 3] + (25, 9\%) C[2, 2] + (55, 9\%) C[2, 3].$$

5.6 The centroid of the face

The purpose of this section is to compute the "average face". Note that we cannot just compute the average of the matrices for the three images on page 143 for example. The images could be of different sizes. Furthermore, the faces could be at different locations in the image. We need to determine the location of the face in each image. To do so, we use eight landmarks. We identify the location of the inner and outer eye, the nostril, and the outer mouth, both on the right and the left. Each landmark is located at a certain row (height) and column (width).

We define the centroid of the face as being the point $(\bar{c}, \bar{r})$, where $\bar{c}$ and $\bar{r}$ are respectively the averages of the eight columns and the eight rows of the landmarks. In Figure 5.4, the landmarks are identified with x's and the centroid by a small circle.

To compute the average of our three faces in the images on page 143, for each image, we keep an image of size 200 pixels by 200 pixels centered about the centroid of the face. Each image is represented by a matrix. We compute the average of the three matrices. The result is in Figure 5.6 on page 154. The original images are in the diagonal of the array. The images off the diagonal in the upper right corner are pairwise averages. The average of the three images is in the lower corner on the left.

We should notice that finding the center of the face is not sufficient to make an average face. The eyes in image 1 (at Tremblant) are slanted compared to the eyes in image 2 (at Vancouver). Furthermore, image 3 (at Morocco) appears to be on a different scale than the other images. The face appears to be closer to the camera in this last image. We will have to transform the images to try to align the landmarks, to get the eyes on the

Fig. 5.4 Eight landmarks to find the centroid.

Fig. 5.5 Landmarks are vectors with respect to the centroid.

eyes, and so on.

5.7 Optimal transformation for the average face

In this section, we discuss the transformation of the images to try to align the landmarks. We refer to a point (x, y) in the image by using the coordinates $[x - \bar{c}, y - \bar{r}]$ with respect to the centroid. In Figure 5.5, we are displaying each landmark as a vector. As you can imagine, it is important that a transformation does not change the shape of the face. To this end, we use a uniform scaling and a rotation about the centroid of the face.

To find the "optimal" transformation to align the faces, we need to define a measure of distance between the two faces. Since a landmark is a vector, we could define the distance between corresponding landmarks as the Euclidian distance between vectors. Consider two vectors in $\mathbb{R}^2$, say $\vec{v} = [v_x, v_y]$ and $\vec{u} = [u_x, u_y]$. The Euclidian distance between $\vec{v}$ and $\vec{u}$ is

$$d(\vec{v}, \vec{u}) = \sqrt{(v_x - u_x)^2 + (v_y - u_y)^2}.$$

For example, the vector corresponding to the exterior right eye is $[-25, -8]$ for Tremblant, but it is $[-23.5, -9.5]$ in Vancouver. So the distance between these corresponding landmarks is $\sqrt{(-25 - (-23.5))^2 + (-9.5 - (-8))^2} = 2.121$.

To get a total distance (in square units) between the faces, we square the Euclidian distance between the two landmarks and compute the sum of the squared distances over the eight landmarks. For our three images, we compute the total distance (in square units):

$$Q(\text{Tremblant}; \text{Vancouver}) = 56,$$

$$Q(\text{Tremblant}; \text{Morocco}) = 297,$$

$$Q(\text{Vancouver}; \text{Morocco}) = 266.$$

Since the face in Vancouver is the closest to the other images, we try to align the other two images to it.

Consider the transformation of the face in the image at Tremblant. We want it to be as "close" as possible to the face in the image in Vancouver. We can try to find a uniform scaling and rotation that will give a distance of zero, but in practice such a transformation will not exist. Why? The image is a planar cross-section of space and the images in different photos usually do not correspond exactly to the same cross-section. So the best we can do is to try to make the distance as small as possible.

Fig. 5.6   Averaging centered faces.

angle = 6.8 degrees
factor = 98.8%

angle = 2.8 degrees
factor = 77.7%

Fig. 5.7   Optimal transformation of the faces.

We transform the landmarks in Tremblant with a uniform scaling and a rotation about the centroid of the face. The matrix of the transformation is of the form $A_T = \begin{bmatrix} a & -b \\ b & a \end{bmatrix}$, where a and b are real numbers. The vector for the exterior right eye is $[-25, -8]$ for Tremblant. After applying the transformation, the vector becomes

$$\begin{bmatrix} a & -b \\ b & a \end{bmatrix} \begin{bmatrix} -25 \\ -8 \end{bmatrix} = \begin{bmatrix} -25\,a + 8\,b \\ -25\,b - 8\,a \end{bmatrix}.$$

We want this vector to correspond to the vector for the exterior right eye in Vancouver, which is $[-23.5, -9.5]$. So we get the following system of linear equations in (a, b):

$$-23.5 = -25\,a + 8\,b$$
$$-9.5 = -25\,b - 8\,a.$$

Repeating for the other seven landmarks, we get a system of 16 linear equations with two unknowns. Since we will not be able to get the landmarks

Fig. 5.8   Averaging of the faces after alignment.

to align exactly, the system will be inconsistent. However, we will be able to find a solution to this system of equations with the least squares method described in Section 5.9. The solution will align the landmarks as "close" as possible.

Once we have the values for a and b, we can transform the image. We can also use Theorem 5.1 to interpret the transformation in terms of a rotation and a uniform scaling. The optimal transformations are described in Figure 5.7.

Now that the landmarks are aligned (or at least they are as "close" as possible by using the least-squares method), we are ready to construct the average. The average of the three aligned faces is in position $(3, 1)$ in Figure 5.8. The image at position $(1, 2)$ in Figure 5.8 is the average of the diagonal images in positions $(1,1)$ and $(2,2)$. The image at position $(1,3)$ in Figure 5.8 is the average of the diagonal images in positions $(1,1)$ and $(3,3)$. The image at position $(2,3)$ in Figure 5.8 is the average of the diagonal images in positions $(2,2)$ and $(3,3)$. Compare the averages in Figure 5.6 to those in Figure 5.8. By using the least-square method, we are able to get much better result.

5.8 Convex sets and extremal points

In this section, we study the concept of convex sets and extremal points.

Definition 5.1.

(i) A subset K of $\mathbb{R}^2$ is called a **convex set** if for every pair of vectors $\vec{v}_1$ and $\vec{v}_2$ in K, we have

$$t\,\vec{v}_1 + (1-t)\vec{v}_2 \in K,$$

for every $t \in [0,1]$. We say that the vector $t\,\vec{v}_1 + (1-t)\vec{v}_2$ is a convex combination of $\vec{v}_1$ and $\vec{v}_2$.

(ii) Let K be a convex set and $\vec{v}$ an element of K. We say that $\vec{v}$ is an extremal point, if $K \setminus \{\vec{v}\}$ (i.e. K without $\vec{v}$) is also a convex set.

We can think of convex combinations as all the vectors that lie on the "line segment" between $\vec{v}_1$ and $\vec{v}_2$. Refer to the Figure 5.9 for an example of a convex combination of vectors. It should be evident that a rectangular image is a convex set in $\mathbb{R}^2$. Furthermore, the four vertices of the rectangle are the only extremal points in the set.

Given a linear map $T : \mathbb{R}^2 \to \mathbb{R}^2$ and a convex set $K \subseteq \mathbb{R}^2$, is the image $T(K)$ of K a convex set? Moreover, if K is convex and $\vec{v}$ is an extremal point of K, is $T(\vec{v})$ an extremal point of $T(K)$? The following theorem gives the answer with the assumption that T is invertible.

Theorem 5.2. Let $T : \mathbb{R}^2 \to \mathbb{R}^2$ be a linear map and $K \subseteq \mathbb{R}^2$ a convex set.

(a) The image $T(K) = \{T(\vec{v}) : \vec{v} \in K\}$ of K is a convex set.
(b) Assume T is invertible. Then, $\vec{v} \in K$ is an extremal point if and only if $T(\vec{v})$ is an extremal point of $T(K)$.

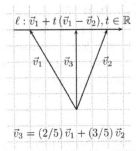

$$\ell : \vec{v}_1 + t\,(\vec{v}_1 - \vec{v}_2), t \in \mathbb{R}$$

$$\vec{v}_1 \quad \vec{v}_3 \quad \vec{v}_2$$

$$\vec{v}_3 = (2/5)\,\vec{v}_1 + (3/5)\,\vec{v}_2$$

Fig. 5.9   Convex combination of two vectors.

Proof.

(a) Let $\vec{v}_1, \vec{v}_2 \in T(K)$, and write $\vec{v}_1 = T(\vec{u}_1), \vec{v}_2 = T(\vec{u}_2)$ for some vectors $\vec{u}_1, \vec{u}_2 \in K$. For any $t \in [0,1]$, we have

$$t\,\vec{v}_1 + (1-t)\,\vec{v}_2 = t\,T(\vec{u}_1) + (1-t)\,T(\vec{u}_2)$$
$$= T(t\,\vec{u}_1 + (1-t)\,\vec{u}_2) \text{ (by the linearity of } T).$$

Since K is convex, $t\,\vec{u}_1 + (1-t)\,\vec{u}_2 \in K$. We conclude that $t\,\vec{v}_1 + (1-t)\,\vec{v}_2 = T(\vec{u})$ for some $\vec{u} \in K$ and therefore $t\,\vec{v}_1 + (1-t)\,\vec{v}_2 \in T(K)$. This shows that $T(K)$ is convex.

(b) Assume for this part that T is invertible. That means that T is a bijective map (injective and surjective) in addition of being linear. Let $\vec{v} \in K$ be any point. We claim that $T(K)\backslash\{T(\vec{v})\} = T(K\backslash\{\vec{v}\})$. To see this, let $\vec{w} \in T(K)\backslash\{T(\vec{v})\}$, then $\vec{w} \in T(K)$ but $\vec{w} \neq T(\vec{v})$. So $\vec{w} = T(\vec{u})$ for some $\vec{u} \in K$ and $\vec{w} = T(\vec{u}) \neq T(\vec{v})$. Since T is bijective, the latter inequality implies that $\vec{u} \neq \vec{v}$. Therefore, $\vec{w} = T(\vec{u})$ with $\vec{u} \in K \backslash \{\vec{v}\}$, so $\vec{w} \in T(K \backslash \{\vec{v}\})$. This shows that $T(K)\backslash\{T(\vec{v})\} \subseteq T(K \backslash \{\vec{v}\})$. The other inclusion is proven similarly.

If $\vec{v} \in K$ is extremal, we need to show that $T(\vec{v})$ is also an extremal point of $T(K)$. To see this, let $\vec{w}_1, \vec{w}_2 \in T(K)\backslash\{T(\vec{v})\}$ and let $t \in [0,1]$. Then, $\vec{w}_1 = T(\vec{u}_1), \vec{w}_2 = T(\vec{u}_2)$ with $\vec{u}_1, \vec{u}_2 \in K \backslash \{\vec{v}\}$ and we have $t\,\vec{w}_1 + (1-t)\,\vec{w}_2 = T(t\,\vec{u}_1 + (1-t)\,\vec{u}_2)$. Since $\vec{v}$ is an extremal point of K, $t\,\vec{u}_1 + (1-t)\,\vec{u}_2 \in K \backslash \{\vec{v}\}$. We conclude that $t\,\vec{w}_1 + (1-t)\,\vec{w}_2 \in T(K \backslash \{\vec{v}\}) = T(K) \backslash \{T(\vec{v})\}$. This proves that $T(\vec{v})$ is an extremal point of $T(K)$. The converse is proved similarly. $\square$

5.9 Least squares method

In this section, we discuss the least squares method to find the "best" solution to an inconsistent system of linear equations. For example, it is possible that we cannot find a transformation that aligns the image perfectly. In this case, we will have to content ourselves with finding the transformation that minimizes the distance between landmarks.

Here is the general setting. Given a system of n linear equations in p variables $\beta_1, \beta_2, \ldots, \beta_p$:

$$\begin{cases} y_1 = \beta_1\, x_{11} + \beta_2\, x_{12} + \cdots + \beta_p\, x_{1p} \\ y_2 = \beta_1\, x_{21} + \beta_2\, x_{22} + \cdots + \beta_p\, x_{2p} \\ \vdots \quad \vdots \qquad\qquad\quad \vdots \\ y_n = \beta_1\, x_{n1} + \beta_2\, x_{n2} + \cdots + \beta_p\, x_{np} \end{cases}$$

we start by writing the system in matrix form: $Y = X\,\beta$, where

$$Y = \begin{bmatrix} y_1 \\ y_2 \\ \vdots \\ y_n \end{bmatrix}, \quad \beta = \begin{bmatrix} \beta_1 \\ \beta_2 \\ \vdots \\ \beta_p \end{bmatrix}, \quad \text{and} \quad X = \begin{bmatrix} x_{11} & x_{12} & \cdots & x_{1p} \\ x_{21} & x_{22} & \cdots & x_{2p} \\ \vdots & \vdots & \ddots & \vdots \\ x_{n1} & x_{n2} & \cdots & x_{np} \end{bmatrix}.$$

We are interested in finding a vector $\hat{\beta} \in \mathbb{R}^p$, such that $X\,\hat{\beta}$ is as "close" as possible to Y. Of course, we have to clearly define what we mean by "close" in this setting. We start by reviewing some notions from linear algebra.

5.9.1 *Dot product - Inner product*

Definition 5.2. Let

$$v = \begin{bmatrix} v_1 \\ v_2 \\ \vdots \\ v_n \end{bmatrix} \quad \text{and} \quad w = \begin{bmatrix} w_1 \\ w_2 \\ \vdots \\ w_n \end{bmatrix}$$

be two vectors in $\mathbb{R}^n$. The dot product of v and w is defined as the scalar

$$v \cdot w = \sum_{i=1}^{n} v_i\, w_i = v^t\, w. \tag{5.2}$$

The last expression in (5.2) gives the dot product in terms of matrix multiplication. This means in particular that the dot product inherits many of its properties from matrix multiplication. Some of these properties are listed in the following theorem.

Theorem 5.3. Let u, v and w be vectors in $\mathbb{R}^n$ and let $c \in \mathbb{R}$. The dot product satisfies the following properties.

(i) $v \cdot w = w \cdot v$ (commutativity of the dot product)

(ii) $(u + v) \cdot w = (u \cdot v) + (v \cdot w)$ (distributivity of the dot product with respect to addition)

(iii) $(cv) \cdot w = c(v \cdot w) = v \cdot (cw)$

(iv) $v \cdot v \geq 0$

(v) $v \cdot v = 0$ if and only if $v = 0$.

Proof.

(i) Write $v = \begin{bmatrix} v_1 \\ v_2 \\ \vdots \\ v_n \end{bmatrix}$, and $w = \begin{bmatrix} w_1 \\ w_2 \\ \vdots \\ w_n \end{bmatrix}$. Then, $v \cdot w = \sum_{i=1}^{n} v_i w_i = \sum_{i=1}^{n} w_i v_i = w \cdot v$.

(ii) $(u + v) \cdot w = (u + v)^t w = (u^t + v^t) w = u^t w + v^t w = (u \cdot w) + (v \cdot w)$.

(iii) $(cv) \cdot w = (cv)^t w = (cv^t) w = c(v^t w) = c(v \cdot w)$.

(iv) Write $v = \begin{bmatrix} v_1 \\ v_2 \\ \vdots \\ v_n \end{bmatrix}$, then

$$v \cdot v = \sum_{i=1}^{n} v_i^2 \geq 0.$$

(v) If $v = 0$, then $v_i = 0$ for $i = 1, \ldots, n$, where $v = \begin{bmatrix} v_1 \\ v_2 \\ \vdots \\ v_n \end{bmatrix}$. Therefore,

$0 = \sum_{i=1}^{n} v_i^2 = v \cdot v$. Conversely, if $v \cdot v = 0$, then $\sum_{i=1}^{n} v_i^2 = 0$. Since all the terms in the sum are non-negative, the only way that this sum can equal zero is that $v_i = 0$ for $i = 1, \ldots, n$. Hence, $v = 0$. $\square$

The notion of the dot product of vectors in $\mathbb{R}^n$ is a particular case of a certain type of vector product in a general vector space V, namely the inner product.

Definition 5.3. Let V be a vector space over $\mathbb{R}$. An inner product on V is a function $< \cdot, \cdot >$ that transforms a couple (u, v) of vectors in V into a scalar $< u, v >$ such that for all vectors $u, v, w \in V$ and for every scalar $c \in \mathbb{R}$, the following properties are satisfied:

(i) $< v, w > \; = \; < w, v >$
(ii) $< u + v, w > \; = \; < u, v > + < v, w >$
(iii) $< cv, w > \; = \; c < v, w >$
(iv) $< v, v > \; \geq 0$
(v) $< v, v > \; = \; 0$ if and only if $v = 0$.

Clearly, the dot product defined above in $\mathbb{R}^n$ is an inner product. Recall that we expressed earlier the dot product in $\mathbb{R}^n$ in terms of matrix multiplication, namely $v \cdot w = v^t w$. Using matrix multiplication, we will introduce another inner product in $\mathbb{R}^n$. First, recall some well known notions and facts from Linear Algebra.

For an $n \times p$ matrix X.

- The rank of X (denoted $\text{rank}(X)$) is the number of leading 1's in any echelon form of X.
- The nullity of X (denoted $\dim(\text{Null}(X))$) is the dimension of the subspace

$$\text{Null}(X) = \{v \in \mathbb{R}^p : Xv = 0\} \quad \text{of} \ \mathbb{R}^p.$$

- The rank-nullity theorem states that

$$p = \text{rank}(X) + \dim(\text{Null}(X)).$$

- The following statements are equivalent:

  (i) The columns of X are linearly independent vectors in $\mathbb{R}^n$.
  (ii) The homogeneous system of equations $Xv = 0$ has only the trivial solution $v = 0$.
  (iii) $\text{rank}(X) = p$.
  (iii) $\dim(\text{Null}(X)) = 0$.

We use the above facts to prove the following result.

Theorem 5.4. Let X be an $n \times p$ matrix. If the columns of X are linearly independent, then $X^t X$ is invertible.

Proof. The $p \times p$ matrix is invertible if and only if its rank is equal to p. By the rank-nullity theorem, $\text{rank}(X^t X) = p - \dim(\text{Null}(X^t X))$. It suffices therefore to show that $\text{Null}(X^t X) = \{0\}$.

Since the columns of X are linearly independent, then its nullity is zero (which implies that $\text{Null}(X) = \{0\}$). Let v be a vector in $\text{Null}(X^t X)$. We show that v must be the zero vector. Since, $v \in \text{Null}(X^t X)$, then $X^t X v = 0$. This implies that $v^t X^t X v = v^t 0 = 0$. But, $v^t X^t X v = (X v)^t (X v) = (X v) \cdot (X v)$ (where $u \cdot w$ is the dot product in $\mathbb{R}^n$). By property (v) of the dot product, we conclude that $X v = 0$ and therefore $v \in \text{Null}(X) = \{0\}$. Thus, $v = 0$ and therefore $\text{Null}(X^t X) = \{0\}$ □

Given an $n \times p$ matrix X with $\text{rank}(X) = p$ and two vectors u and v in $\mathbb{R}^p$ written as columns, define

$$< u, v > = u^t (X^t X) v.$$

Notice that the expression $u^t (X^t X) v$ is indeed a scalar.

Theorem 5.5. Given an $n \times p$ matrix X with $\text{rank}(X) = p$, the formula $< u, v > = u^t (X^t X) v$ is an inner product in $\mathbb{R}^p$.

Proof. Let u, v, and w be vectors in $\mathbb{R}^p$ written as columns and let c be a real scalar. Notice that

$$< u, w > = u^t (X^t X) w = (X u)^t (X w) = (X u) \cdot (X w),$$

where $\cdot$ is the dot product on $\mathbb{R}^n$. Both $X u$ and $X w$ are vectors in $\mathbb{R}^n$.

(i) By the commutativity of the dot product, we get $< u, w > = (X u) \cdot (X w) = (X w) \cdot (X u) = < w, u >$.

(ii) $< u + v, w > = (u + v)^t (X^t X) w = (u^t + v^t) (X^t X) w = u^t (X^t X) w + v^t (X^t X) w = < u, w > + < v, w >$.

(iii) $< cv, w > = (cv)^t (X^t X) w = (cv^t) (X^t X) w = c (v^t (X^t X) w)$. The latter is equal to $c < v, w >$. So, $< cv, w > = c < v, w >$.

(iv) $< v, v > = (X v) \cdot (X v) \geq 0$, by property (iv) of the dot product.

(v) Clearly, if $v = 0$, then $< v, v > = 0$. Conversely, let v be a vector in $\mathbb{R}^p$ such that $< v, v > = 0$. So, $0 = < v, v > = v^t (X^t X) v = (X v) \cdot (X v)$. By property (v) of the dot product, we get $X v = 0$. Remember that X was chosen to be of rank p, which means in particular that the homogeneous system of equations $X v = 0$ has a unique solution $v = 0$. Therefore, $< v, v > = 0$ implies that $v = 0$. □

Let us now go back to our original problem in this section. We are interested in finding the vector $\beta \in \mathbb{R}^p$ such that $X\beta$ is as "close" as possible to Y. The Euclidean distance between Y and $X\beta$ is

$$\| Y - X\beta \| = \left[\underbrace{\sum_{i=1}^{n}(y_i - \beta_1\, x_{i1} - \beta_2\, x_{i2} - \cdots - \beta_p\, x_{ip})^2}_{Q(\beta)} \right]^{1/2} = \sqrt{Q(\beta)}.$$

Finding $\hat{\beta} \in \mathbb{R}^p$ such that $X\hat{\beta}$ is "as close as possible" to Y means minimizing the expression $Q(\beta)$. Note that

$$\begin{aligned}
Q(\beta) &= \sum_{i=1}^{n}[y_i - (\beta_1\, x_{i1} + \beta_2\, x_{i2} + \cdots + \beta_p\, x_{ip})]^2 \\
&= (Y - X\,\beta)^t(Y - X\,\beta) \\
&= Y^tY - Y^tX\,\beta - \beta^tX^tY + \beta^tX^tX\beta \\
&= Y^tY - (\beta^tX^tY)^t - \beta^tX^tY + \beta^tX^tX\beta \\
&= Y^tY - 2\,\beta^tX^tY + \beta^tX^tX\beta,
\end{aligned}$$

since β^tX^tY is just a scalar, then it is equal to its transpose. Moreover, if we assume that X is of rank p, the columns of X are linearly independent, and therefore X^tX is invertible by Theorem 5.4. We have

$$\begin{aligned}
Q(\beta) &= Y^tY - 2\,\beta^t(X^tX)(X^tX)^{-1}X^tY + \beta^tX^tX\beta \\
&= Y^tY - 2 < \beta, (X^tX)^{-1}X^tY > + < \beta, \beta > \\
&= Y^tY - 2 < \beta, (X^tX)^{-1}X^tY > + < \beta, \beta > \\
&\quad + < (X^tX)^{-1}X^tY, (X^tX)^{-1}X^tY > \\
&\quad - < (X^tX)^{-1}X^tY, (X^tX)^{-1}X^tY > \\
&= Y^tY + < \beta - (X^tX)^{-1}X^tY, \beta - (X^tX)^{-1}X^tY > \\
&\quad - < (X^tX)^{-1}X^tY, (X^tX)^{-1}X^tY > \\
&= Q_1(\beta) + C,
\end{aligned}$$

where

$$Q_1(\beta) = < \beta - (X^tX)^{-1}X^tY, \beta - (X^tX)^{-1}X^tY >$$

and

$$C = Y^tY - < (X^tX)^{-1}X^tY, (X^tX)^{-1}X^tY > .$$

Note that the expression C is a constant with respect to β. Thus, to minimize $Q(\beta)$ it suffices to minimize $Q_1(\beta)$. Using property (iv) of the inner product, $Q_1(\beta) \geq 0$ for all $\beta \in \mathbb{R}^p$ (since it is of the form $< u, u >$). Therefore, the minimal value for $Q_1(\beta)$ is zero. Using property (v) of the inner product, $Q_1(\beta) = 0$ if and only if $\beta - (X^t X)^{-1} X^t Y = 0$ or equivalently $\beta = (X^t X)^{-1} X^t Y$.

Here is a summary of the least squares method:

- Given a linear system $Y = X\beta$, where X is an $n \times p$ matrix of rank p, the best approximation of a solution to the system is given by the vector

$$\hat{\beta} = (X^t X)^{-1} X^t Y.$$

- Note that if X has fewer rows than columns, then rank$(X) < p$ and the method of least squares will not work.

5.10 Face recognition

In this section, we discuss an algorithm that is used for face recognition. We have an image of the face of an individual suspected of committing a crime and we want to compare this image to images of "average" faces stored in a database. To do so, we measure the distance between the suspect's image and the images in the database. We discuss a measure of distance based on the method of principal components, which is called the "Eigenface" method.

To illustrate the method, assume that our database consists of the six images of penguins in Figure 5.10. As you can imagine, a face is not perfectly the same from one image to another. For example, the lighting can vary from image to image. To simulate a variation in the image, we have distorted the image of first penguin (named Alice) and the fourth penguin (named Bob) from Figure 5.10.

The distorted images of Alice and Bob are found in Figure 5.11. A human eye can clearly identify the images in Figure 5.11 as being those of Alice and Bob, but with a more severe level of distortion and a bigger image database, it would be much harder to identify images using just our human eyes. We explore a method that enables a computer to identify the two distorted images in Figure 5.11 as being Bob and Alice, respectively.

Fig. 5.10   A database of six penguins.

Fig. 5.11   Distorted images of Penguins 1 and 4.

5.10.1   *Descriptive statistics*

5.10.1.1   *At the level of the image*

The images in Figure 5.10 are grayscale raster images, each of size 50×50 pixels. Each image has $p = 50 \times 50 = 2500$ grayscale levels (also called intensities). To describe an image, we compute the mean intensity and the standard deviation of the intensities. We also define a standardized intensity to control varying lighting conditions.

Consider the image of Alice, that is the first penguin in Figure 5.10. We write the corresponding intensities as a column vector $\boldsymbol{x}$:

$$\boldsymbol{x} = \begin{bmatrix} x_1 \\ x_2 \\ \vdots \\ x_{2500} \end{bmatrix} = \begin{bmatrix} 255 \\ 255 \\ \vdots \\ 240 \end{bmatrix}.$$

Recall that these intensities are written in a raster fashion, left to right, top to bottom. The jth component of $\boldsymbol{x}$ (denoted x_j) is the intensity of the jth pixel. The mean intensity is

$$\overline{x} = \frac{\sum_{j=1}^{p} x_j}{p} = \frac{255 + 255 + \cdots + 240}{2500} = 184.2088$$

and the standard deviation of the intensity is

$$\sigma_x = \sqrt{\frac{\sum_{i=1}^{p}(x_i - \overline{x})^2}{p}}$$

$$= \sqrt{\frac{(255 - 184.2088)^2 + (255 - 184.2088)^2 + \cdots + (255 - 184.2088)^2}{2500}}$$

$$= 95.24998.$$

Looking at the expression inside the square root of the standard deviation, we observe that it computes the average squared deviation away from the mean. It represents an average distance away from the mean in squared units. The square root is used to obtain a measure in the same units as the intensities. So the standard deviation is a measurement of variability. The more the intensities are dispersed about the mean, the larger the standard deviation. The more the intensities are concentrated about the mean, the smaller the standard deviation.

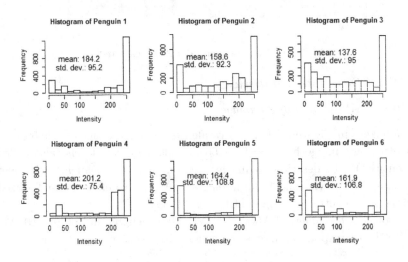

Fig. 5.12   Descriptive statistics of the 6 images of the penguins.

The descriptive statistics for each of the $n = 6$ penguins in our database are displayed in Figure 5.12. The distributions of the intensity are not the same for these six images. Since pictures of faces are taken under varying lighting conditions, the images of faces usually exhibit varying distributions of intensity. We want a system that compares varying characteristics of faces, not varying lighting conditions.

We use a standardized intensity: $z_j = (x_j - \overline{x})/\sigma_x$, for $j = 1, 2, \ldots, p$.

The mean of the standardized intensities is

$$\overline{z} = \frac{1}{p} \sum_{j=1}^{p} \frac{(x_j - \overline{x})}{\sigma_x} = \frac{1}{\sigma_x p} \left[\left(\sum_{j=1}^{p} x_j \right) - p\overline{x} \right] = 0,$$

since $p\overline{x} = \sum_{j=1}^{p} x_j$.

The variance (i.e. the square of the standard deviation) of the standardized intensities is

$$\sigma_z^2 = \frac{1}{p}\sum_{i=1}^{p}(z_i^2 - \overline{z})^2 = \frac{1}{p}\sum_{i=1}^{p}z_i^2 \quad (\text{since } \overline{z} = 0)$$

$$= \frac{1}{p}\sum_{i=1}^{p}\left[(x_j - \overline{x})/\sigma_x\right]^2 = \frac{1}{(p\sigma_x^2)}\sum_{i=1}^{p}(x_j - \overline{x})^2 = \frac{\sigma_x^2}{\sigma_x^2} = 1.$$

By using the standardized intensities to compare images, we have control on varying lighting conditions since all the images would have the same mean and the same standard deviation.

To compare two images, we use the Euclidian distance between their respective vectors of standardized intensities. Let z, w be the respective vectors of the standardized intensities for the two images. We use the following metric (i.e. measure of distance) to compare the two images:

$$\|z - w\| = \sqrt{\sum_{j=1}^{p}(z_j - w_j)^2} = (z - w)^t(z - w).$$

Bearing in mind that someone's face can vary from image to image, this means that the image we are comparing to the database will not be exactly the same as any of the images in the database. So we will need to rank them from the closest to our image to the furthest away.

We compare the distorted images of Alice and Bob (see Figure 5.11) to our database of six penguins (see Figure 5.10). The distances between the images are found in Table 5.1. Among the six penguins in the database, it is the image of Alice that is closest to the distorted image of Alice. However, the distance of 14 between the two images of Alice is not considered very small.

In practice, an arbitrary threshold ϵ is used as a rule to determine if the person in the image is in the database. If ϵ was set to 10, then the computer would conclude that the penguin in the distorted image of Alice is not in the database. The conclusion would be similar for Bob, since the distorted image of Bob is more than 10 units away from all of the images in the database.

The root of the problem is that we are comparing all of the $p = 2500$ pixels. We are using some information in the comparison that is not relevant to the differences between the 6 penguins in the database. We need to describe the images at the level of the pixels to identify variations between the images of the penguins.

Table 5.1 Distances between the images of the penguins.

Penguin	1	2	3	4	5	6
			Penguin			
Distorted Alice	14.0	54.8	62.5	47.5	52.1	49.0
Distorted Bob	47.1	52.7	61.9	11.2	59.0	52.9
1	0	54.3	62.0	46.8	51.7	48.3
2		0	57.1	52.8	66.0	60.9
3			0	62.0	64.7	61.3
4				0	58.9	53.3
5					0	39.9
6						0

5.10.1.2 *At the level of a pixel*

Let $z_1, z_2, \ldots, z_n$ be the respective vectors of the standardized intensities of the $n = 6$ images in the database. They correspond to the 6 penguins in Figure 5.10. Consider these vectors as the rows of an $n \times p$ matrix (called a *data array* by statisticians):

$$Z = \begin{bmatrix} z_1 \\ z_2 \\ \vdots \\ z_n \end{bmatrix} = \begin{bmatrix} c_1 & c_2 & \cdots & c_p \end{bmatrix}. \tag{5.3}$$

The jth column (that is $c_j \in \mathbb{R}^n$) of the data array Z is the vector of the standardized intensities of the jth pixel from the $n = 6$ images.

If the intensities between the images are very dispersed (i.e. there is a large variance) at a particular pixel, then this pixel will be considered important in the differentiation of the images. Furthermore, if two pixels are highly correlated, then we only need one of these two pixels (or we might consider combining the information from the two pixels), since they contain similar information.

As an example, let us consider the 1000th pixel and the 1500th pixel (each has been rounded to three decimal places) from our database of 6 penguins:

$$u = c_{1000} = \begin{bmatrix} -0.979 \\ 0.189 \\ -0.711 \\ 0.249 \\ 0.300 \\ -0.917 \end{bmatrix}, \quad v = c_{1500} = \begin{bmatrix} -0.475 \\ 0.135 \\ -0.669 \\ 0.223 \\ 0.005 \\ -0.757 \end{bmatrix}.$$

The variance of the (standardized) intensities for the 1000th pixel and the 1500th pixel are respectively

$$\sigma^2_{1000} = \frac{1}{n}\sum_{i=1}^{n}(u_i - \bar{u})^2 = 0.318 \text{ and } \sigma^2_{1500} = \frac{1}{n}\sum_{i=1}^{n}(v_i - \bar{v})^2 = 0.153.$$

Among these two pixels, the 1000th is more variable in terms of the (standardized) intensities. The larger the variance, the more this pixel will be useful in the differentiation of the images. So if we were to retain only one of these two pixels for the comparison of the images, we should choose the 1000th pixel. However, we will see that combining the information from the two pixels will give us a component that is even more variable.

Before discussing the combination of pixels, we look at the statistical association between these two pixels. A scatter plot of the 1500th pixel against the 1000th pixel is found in Figure 5.13. The vertical and the horizontal lines in the middle of the plot are the respective means of the two variables. We see that they are highly associated. When one of the pixels is large compared to its mean, so is the other pixel. Furthermore, when one of the pixels is small compared to its mean, so is the other pixel. This type of association is called positive correlation. When the majority of the points are in Quadrants I and III, we say the association is positive. While a negative association occurs when the majority of the points are in Quadrants II and IV.

A statistic that captures the sign (i.e. positive or negative) of the association is the covariance. The covariance is computed as follows:

$$\sigma_{1000,1500} = \frac{1}{n}\sum_{i=1}^{n}(u_i - \bar{u})(v_i - \bar{v}) = 0.2064.$$

If there is no association between the pixels, then the points in the scatter plot are going to be scattered in all four quadrants delimited by the respective means of these two variables. In this case, the covariance should be close to zero.

If the association between the pixels is positive, then the majority of the points are in Quadrants I and III. Thus, the product $(u_i - \bar{u})(v_i - \bar{v})$ is positive for the majority of the points, which gives a positive covariance. If the association between the pixels is negative, then the majority of the points are in Quadrants II and IV. Thus, the product $(u_i - \bar{u})(v_i - \bar{v})$ is negative for the majority of the points, which gives a negative covariance.

To measure the intensity of the association, we compute the correlation as follows:

$$\rho_{1000,1500} = \frac{\sigma_{1000,1500}}{\sigma_{1000}\,\sigma_{1500}} = 0.934.$$

If the pixels are not associated, then the correlation should be near zero, since the covariance will be near zero. We recall a theorem from linear algebra to help us interpret a non-zero correlation.

Theorem 5.6. Let $x = [x_1, x_2, \ldots, x_n]^t$ and $y = [y_1, y_2, \ldots, y_n]^t$ be vectors in $\mathbb{R}^n$. Then

$$-\sqrt{\sum_{i=1}^{n} x_i^2}\sqrt{\sum_{i=1}^{n} y_i^2} \leq \sum_{i=1}^{n} x_i\, y_i \leq \sqrt{\sum_{i=1}^{n} x_i^2}\sqrt{\sum_{i=1}^{n} y_i^2}.$$

These inequalities are called the *Cauchy-Schwarz inequalities*. Furthermore, the inequalities are strict, unless one of the vectors x, y is a multiple of the other.

Taking $x_i = u_i - \overline{u}$ and $y_i = v_i - \overline{v}$ for $i = 1, \ldots, n$ in the Cauchy-Schwarz inequalities, we get

$$-1 \leq \rho_{1000,1500} \leq 1$$

and the inequalities are strict unless $u_i = \overline{u} + c\,(v_i - \overline{v})$ for some c, for all $i = 1, 2, \ldots, n$ or $v_i = \overline{v} + c\,(u_i - \overline{u})$ for some c, for all $i = 1, 2, \ldots, n$. This means that a correlation is always between -1 and 1. Furthermore, it is equal to 1 or -1 only if the points in the scatter plot fall exactly on a line. Falling exactly on a line is a very strong association, that is called **perfect correlation**.

We interpret a correlation closer to 1 or -1 as a more intense association. The correlation between the 1000th pixel and the 1500th pixel is 0.934. This means that they are highly correlated.

To combine the information found in these two pixels, we orthogonally project the points in the scatter plot onto a line that passes through the centroid $(\overline{u}, \overline{v}) = (-0.311, -0.256)$. For our example, we choose two lines. The line 1 will pass through Quadrants I and III, i.e. in the part of the plane that corresponds to positive correlation. This line will have the vector $\vec{d_1} = (-0.8279119, -0.5608582)$ as unit directional vector. The projected points are the triangles in Figure 5.14. We use the values along this line for our two pixels. To compute the scalar projection of the ith point, we extend a vector $\vec{y_i}$ from the centroid to the point. The scalar projection is

$w_i = \vec{y}_i \cdot \vec{d}_1$, where $\cdot$ is the dot product in $\mathbb{R}^2$. As an example, consider the point $(-0.979, -0.475)$. Its corresponding vector is $\vec{y} = (-0.979, -0.475) - (\overline{u}, \overline{v}) = (-0.668, -0.219)$ and the corresponding scalar projection onto the line is

$$w = \vec{y} \cdot \vec{d}_1 = (-0.668)(-0.8279119) + (-0.219)(-0.5608582) = 0.675.$$

The 6 scalar projections along the line with direction vector $\vec{d}_1$ are

$$w_1 = 0.675, \quad w_2 = -0.633, \quad w_3 = 0.562,$$

$$w_4 = -0.733, \quad w_5 = -0.652, \quad w_6 = 0.782.$$

Interpret these values as deviations away from the center $(\overline{u}, \overline{v}) = (-0.311, -0.256)$ along the line with the direction vector $\vec{d}_1$. The variance of these values is

$$s_w^2 = \frac{1}{n} \sum_{i=1}^{n} (w_i - \overline{w})^2 = 0.550.$$

The combined component is more variable than the 1000th pixel and the 1500th pixel. Recall that $\sigma_{1000}^2 = 0.318$ and $\sigma_{1500}^2 = 0.153$. So the combined component will be more useful in the differentiation of the images compared to the 1000th pixel or to the 1500th pixel.

We do have to be careful in the combination of the pixels, since we can obtain a component that has a smaller variance than the original pixels. If we project the vectors onto the line 2 in Figure 5.14, which has the direction vector $\vec{d}_2 = (0.5608582, -0.8279119)$, we get

$$w_1 = -0.193, \quad w_2 = -0.043, \quad w_3 = 0.117,$$

$$w_4 = -0.082, \quad w_5 = 0.126, \quad w_6 = 0.075.$$

The variance of these values is $s_w^2 = 0.0162$. This is a reduction of variance compared with each of the original pixels. Combining the pixels does not always give more variance. Clearly the values are more varied along line 1 compared to the values along line 2. In fact, the points along line 1 resemble much more the original points compared to the points along line 2. The correlation of the points along line 2 is negative, while the correlation of the 6 original points is positive.

In the next section, we learn how to combine the information found in all $p = 2500$ pixels by using a covariance matrix that will contain all variances and covariances.

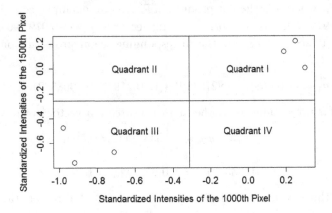

Fig. 5.13   Scatter plot of the 1500th pixel versus the 1000th pixel.

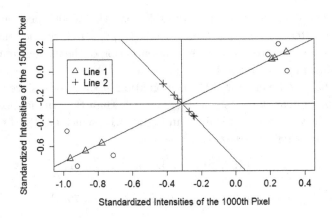

Fig. 5.14   Projections of the points onto lines passing through the centroid.

5.10.2   *The principle components from the covariance matrix*

Our illustrative example in the previous subsection only describes the combination of two pixels, that is the 1000th against the 1500th pixel. However, we want to combine all pixels. To do so, we construct the covariance matrix (of size $p \times p$):

$$V = \begin{bmatrix} \sigma_1^2 & \sigma_{12} & \dots & \sigma_{1p} \\ \sigma_{21} & \sigma_2^2 & \dots & \sigma_{2p} \\ \vdots & & \ddots & \vdots \\ \sigma_{p1} & \sigma_{p2} & \dots & \sigma_p^2 \end{bmatrix}.$$

The jth element in the diagonal, i.e. σ_j^2, is the variance of the standardized intensities of the jth pixel. The off-diagonal element σ_{jk} is the covariance between the standardized intensities of the jth pixel and the standardized intensities of the kth pixel.

We can compute V from the rows of the data array Z as defined in (5.3). The ith row $\boldsymbol{Z}_i$ of the data array Z contains the $p = 2500$ standardized intensities for the ith image in the data base. Compute the mean vector of the rows in $\boldsymbol{Z}$:

$$\overline{\boldsymbol{Z}} = \frac{1}{n} \sum_{i=1}^{n} \boldsymbol{Z}_i.$$

The jth component of $\overline{\boldsymbol{Z}}$ is the mean of the jth pixel. We want to combine the pixels to form one component by projecting $\boldsymbol{Z}_i$ onto a line that passes through the centroid $\overline{\boldsymbol{Z}}$. To do so, we define the column vector $\boldsymbol{Y}_i$ with the centroid $\overline{\boldsymbol{Z}}$ as the initial point and $\boldsymbol{Z}_i$ as the terminal point, that is

$$\boldsymbol{y}_i = \boldsymbol{Z}_i^t - \overline{\boldsymbol{Z}}^t.$$

The mean of the vectors $\boldsymbol{y}_i$ is the zero vector $\boldsymbol{0}$,

$$\frac{1}{n} \sum_{i=1}^{n} (\boldsymbol{Z}_i^t - \overline{\boldsymbol{Z}}^t) = \frac{1}{n} \left(\sum_{i=1}^{n} \boldsymbol{Z}_i^t \right) - \overline{\boldsymbol{Z}}^t = \overline{\boldsymbol{Z}}^t - \overline{\boldsymbol{Z}}^t = \boldsymbol{0}.$$

The following representation of the covariance matrix V in terms of the vectors $\boldsymbol{y}_i$ will be useful:

$$V = \frac{1}{n} \sum_{i=1}^{n} \boldsymbol{y}_i \boldsymbol{y}_i^t. \tag{5.4}$$

To show that the latter matrix is indeed the covariance matrix, consider $\boldsymbol{u} = (u_1, u_2, \dots, u_n)$ and $\boldsymbol{v} = (v_1, v_2, \dots, v_n)$ to be the standardized intensities of the jth and the kth pixel, respectively. The vectors $\boldsymbol{u}$ and $\boldsymbol{v}$ are the jth and the kth columns of the data array Z. The (j, k) element of V is the covariance between u and v. The (j, k) element of the matrix $\boldsymbol{y}_i \boldsymbol{y}_i^t$ is $(u_i - \overline{u})(v_i - \overline{v})$. This means that the (j, k) element of $\frac{1}{n} \sum_{i=1}^{n} \boldsymbol{y}_i \boldsymbol{y}_i^t$ is

$$\frac{1}{n} \sum_{i=1}^{n} (u_i - \overline{u})(v_i - \overline{v}).$$

The latter is the covariance between u and v. Thus, $\frac{1}{n}\sum_{i=1}^{n}y_i y_i^t$ is the covariance matrix V.

To combine the pixels, we orthogonally project the point $z_i \in \mathbb{R}^p$ (which corresponds to the ith image) onto a line in $\mathbb{R}^p$ that passes through the centroid $\overline{Z}$ and with direction vector $e \in \mathbb{R}^p$. We assume that the direction vector is a unit vector e (that is $\|e\| = 1$ or equivalently $e^t e = 1$). The corresponding scalar projection is $y_i \cdot e = y_i^t e$. The scalar projections of the $n = 6$ images are:

$$w_1 = y_1^t e, w_2 = y_2^t e, \ldots, w_n = y_n^t e.$$

The mean of the scalar projections is always zero:

$$\overline{w} = \sum_{i=1}^{n} w_i = \sum_{i=1}^{n} y_i^t e = \left(\sum_{i=1}^{n} y_i^t \right) e = 0^t e = 0.$$

The variance of the scalar projections can be computed from the covariance matrix V:

$$\sigma_w^2 = \frac{1}{n} \sum_{i=1}^{2} (w_i - \overline{w})^2 = (1/n) \sum_{i=1}^{2} w_i^2 \quad (\text{since } \overline{w} = 0)$$

$$= \frac{1}{n} \sum_{i=1}^{2} w_i^t w_i \quad (\text{since } w_i \text{ is a scalar})$$

$$= \frac{1}{n} \sum_{i=1}^{2} (y_i^t e)^t (y_i^t e)$$

$$= \frac{1}{n} \sum_{i=1}^{2} e^t y_i y_i^t e = e^t \left[\frac{1}{n} \sum_{i=1}^{2} y_i y_i^t \right] e = e^t V e.$$

We want to find the direction e that maximizes the variance σ_w^2. The following theorem will be useful in the interpretation of the optimal direction. We state the theorem without proof. It is a well-known result in linear algebra. It states that a symmetric matrix (i.e. the matrix is equal to its transpose) is diagonalizable. The notions of eigenvalues, eigenvectors and orthogonal bases are quickly reviewed in the remarks following the theorem.

Theorem 5.7 (Principal Axis Theorem). Let V be a $p \times p$ real valued symmetric matrix. Then, the p eigenvalues of V are real, and there exists an orthonormal basis of $\mathbb{R}^p$ consisting of eigenvectors of V. Write the eigenvalues in descending order $\lambda_1 \geq \lambda_2 \geq \cdots \geq \lambda_p$ and let e_i be an eigenvector associated to the eigenvalue λ_i for $j = 1, 2, \ldots, p$, such that

$\mathcal{B} = \{e_1, e_2, \ldots, e_p\}$ is an orthonormal basis of $\mathbb{R}^p$. Let P be the matrix, whose columns are the vectors in $\mathcal{B}$. The matrix V has the following decomposition:

$$V = P D P^t = \sum_{j=1}^{p} \lambda_j\, e_j e_j^t.$$

Remarks:

(a) We say that a non-zero vector $e \in \mathbb{R}^p$ is an eigenvector of V, if there exists a scalar λ such that

$$V e = \lambda e.$$

The scalar λ is called the corresponding eigenvalue.

(b) We say that $\mathcal{B} = \{e_1, e_2, \ldots, e_p\}$ is an orthonormal basis of $\mathbb{R}^p$, if

   (i) the vectors in $\mathcal{B}$ are orthogonal, i.e. $e_j^t e_k = 0$ for all $j \neq k$.

   (ii) the vectors in $\mathcal{B}$ are unit vectors, i.e. $\|e_j\| = 1$ for all j. Equivalently this means $e_j^t e_j = 1$ for all j.

(c) As a consequence of part (b), the product $P^t P$ is equal to the identity matrix I, since the columns of P form an orthonormal basis of $\mathbb{R}^p$.

(d) The scalar projection of the ith centered observation $y_i = Z_i^t - \overline{Z}^t$ onto the jth eigenvector e_j, that is $y_i \cdot e_j = y_i^t e_j$, is called the jth **principal component** by statisticians.

Here are a few consequences of the Principle Axis Theorem.

(1) The eigenvalues of the covariance matrix V can be interpreted as variances. Let w be the component corresponding to the scalar projections of the centered observations y_i for $i = 1, 2, \ldots, n$ along the line that passes through the centroid with direction vector v, where v is an eigenvector of V with corresponding eigenvalue λ. As seen above the variance of w is $v^t V v$, which becomes

$$s_w^2 = v^t V v = v^t \left(\lambda v^t\right) = \lambda \left(v^t v\right) = \lambda,$$

since $v^t v = 1$.

(2) The trace of the covariance matrix V is the sum of the elements in its diagonal, that is $\operatorname{tr}(V) = \sum_{j=1}^{p} \sigma_j^2$. It is called the **total variance**. For our database of $n = 6$ penguins, the total variance is $\operatorname{tr}(V) = 1327.045$. We can compute the trace from the sum of the eigenvalues:

$$\operatorname{tr}(V) = \operatorname{tr}(P D P^t) = \operatorname{tr}(P^t P D) = \operatorname{tr}(I D) = \operatorname{tr}(D) = \sum_{j=1}^{p} \lambda_j.$$

We used the fact that the $P^t P = I$, where I is the $p \times p$ identity matrix, and also the cyclic property of the trace: $\text{tr}(A B) = \text{tr}(B A)$ for any matrices A, B such that $A B$ and $B A$ are defined.

Given a symmetric $p \times p$ matrix $A = [a_{ij}]$ and a vector $e = [e_1, e_2, \ldots, e_p]^t \in \mathbb{R}^p$, we define a linear form and a quadratic form of e as follows. For any $b = [b_1, b_2, \ldots, b_p]^t \in \mathbb{R}^p$, the scalar

$$L = b^t A e = e^t A b = \sum_{i=1}^{p} \sum_{k=1}^{p} b_k a_{ki} e_i$$

is called a *linear form* of e. The scalar

$$Q = e^t A e = \sum_{j=1}^{p} \sum_{k=1}^{p} a_{jk} e_j e_k$$

is called a *quadratic form* of e.

Both the linear form L and the quadratic form Q are functions of p variables $e_1, e_2, \ldots, e_p$ (the components of the vector $e \in \mathbb{R}^p$) and one can define the gradient vector for L and Q as being

$$\frac{\partial L}{\partial e} = \begin{bmatrix} \partial L/\partial e_1 \\ \partial L/\partial e_2 \\ \vdots \\ \partial L/\partial e_p \end{bmatrix} \quad \text{and} \quad \frac{\partial Q}{\partial e} = \begin{bmatrix} \partial Q/\partial e_1 \\ \partial Q/\partial e_2 \\ \vdots \\ \partial Q/\partial e_p \end{bmatrix},$$

where as usual $\frac{\partial F}{\partial x_i}$ means the partial derivative of the function F with respect to the variable x_i. The following Theorem gives an easy way to compute the gradient vector.

Theorem 5.8. Let A be a symmetric $p \times p$ matrix, $e = [e_1, e_2, \ldots, e_p]^t$ and $b = [b_1, b_2, \ldots, b_p]^t$ be 2 vectors in $\mathbb{R}^p$. Let L and Q be the linear and quadratic forms as above. Then,

$$\frac{\partial L}{\partial e} = A b \quad \text{and} \quad \frac{\partial Q}{\partial e} = 2 A e.$$

The concepts of linear and quadratic forms in the above theorem are a generalization of first and second order power functions in $\mathbb{R}$. Notice that the formula for the gradient vector (i.e. the vector of derivatives) is an analogue of the formulae to differentiate the two power functions in $\mathbb{R}$, namely $(d/dx)(c x) = c$ and $(d/dx)(c x^2) = 2 c x$.

To find the optimal direction to maximize the variance, we use the Lagrange method. We want to maximize $s_w^2 = e^t V e$ under the constraint that e is a unit vector in $\mathbb{R}^p$. This constraint is equivalent to $e^t e - 1 = 0$. Define the Lagrangian function:

$$\mathcal{L}(\lambda, e) = e^t V e - \lambda (e^t e - 1) = e^t V e - \lambda (e^t I e - 1),$$

where I the identity matrix of size $p \times p$.

To find the optimal direction e, we need to solve

$$\frac{\partial \mathcal{L}}{\partial e} = 0 \quad \text{and} \quad \frac{\partial \mathcal{L}}{\partial \lambda} = e^t e - 1 = 0.$$

The last equality ensures that the constraint is satisfied. Note that $e \neq 0$, since $e'e = 1$. The former equality gives

$$0 = \frac{\partial \mathcal{L}}{\partial e} = 2 V e - 2 \lambda I e.$$

The latter is equivalent to $V e = \lambda e$. This means, that $e \neq 0$ is an eigenvector of V with corresponding eigenvalue λ.

Since λ_1 is the largest eigenvalue, then the corresponding eigenvector v_1 gives the direction of the line that corresponds to the largest variance. Similarly, λ_p is the smallest eigenvalue, so the corresponding eigenvector v_p gives the direction that corresponds to the smallest variance. For our database of $n = 6$ penguins, $\lambda_1 = 474.0232$. This is the variance of the projection along the line with direction vector v_1. The total variance is $\mathrm{tr}(V) = 1327.045$. So the first principal component accounts for $\lambda_1/\mathrm{tr}(V) = 35.72\%$ of the total variance.

We want the projected images to be similar to the original images, so we should try to recover most of the total variance. Let us try to find a second component with maximum variance that is uncorrelated with the first principal component. The first principal component is in the direction given by v_1. Let e be the direction of the second component. Let u and w be the scalar projections onto the lines that pass through the centroid in the direction given by e and v_1, respectively. So $u_i = y_i^t e$ and $w_i = y_i^t v_1$ for $i = 1, \ldots, n$. Similar to the variance, we will can compute the covariance from the covariance matrix V:

$$\frac{1}{n} \sum_{i=1}^{2} (w_i - \overline{w})(u_i - \overline{u}) = \frac{1}{n} \sum_{i=1}^{2} w_i u_i \quad (\text{since } \overline{w} = 0 = \overline{u})$$

$$= \frac{1}{n} \sum_{i=1}^{2} w_i^t u_i \quad (\text{since } w_i \text{ is a scalar})$$

$$= \frac{1}{n} \sum_{i=1}^{2} (y_i^t v_1)^t (y_i^t e) = \frac{1}{n} \sum_{i=1}^{2} v_1^t y_i y_i^t e = v_1^t \left[\frac{1}{n} \sum_{i=1}^{2} y_i y_i^t \right] e$$

$$= v_1^t V e.$$

We want to maximize $s_u^2 = e^t V e$ under the constraints that e is a unit vector in $\mathbb{R}^p$, that is $e^t e - 1 = 0$, and that the two components are uncorrelated, that is $v_1^t V e = 0$. Define the Lagrangian function:

$$\mathcal{L}(\lambda, \lambda_1, e) = e'Ve - \lambda(e'e - 1) - \lambda_1(v_1'Ve)$$
$$= e'Ve - \lambda(e'Ie - 1) - \lambda_1(v_1'Ie)$$

where I the identity matrix of size $p \times p$.

To find the optimal direction e, we need to solve

$$\frac{\partial \mathcal{L}}{\partial e} = \mathbf{0}, \quad \frac{\partial \mathcal{L}}{\partial \lambda} = e^t e - 1 = 0, \quad \text{and} \quad \frac{\partial \mathcal{L}}{\partial \lambda_1} = v_1^t V e = 0.$$

The last two equalities ensures that the constraints are satisfied. The first equality gives

$$0 = \frac{\partial \mathcal{L}}{\partial e} = 2Ve - 2\lambda Ie - \lambda Vv_1. \tag{5.5}$$

Multiply both sides of the equation by v_1^t on the left. Since $v_1^t 0 = 0$, we get

$$0 = 2v_1^t V e - 2\lambda v_1^t e - \lambda_1 v_1^t V v_1$$
$$= -\lambda_1 v_1^t V v_1,$$

since $v_1^t V e = 0$ and $v_1^t e = 0$. Recall that $v_1^t V v_1$ is the variance of the first component. Assuming that this variance is not zero, then we must have $\lambda_1 = 0$. This means that equation (5.5) becomes

$$0 = 2Ve - 2\lambda Ie,$$

which is equivalent to $Ve = \lambda e$. So $e \neq \mathbf{0}$ is an eigenvector of V with eigenvalue λ and we conclude that the second component is in the direction of the eigenvector e_2 corresponding to the second largest eigenvalue λ_2. Furthermore, λ_2 will correspond to the variance of the second component.

You can probably now guess, that if we want a component of maximum variance that is uncorrelated to the first two components, then we should project the points onto a line in the direction of the eigenvector e_3 corresponding to the third largest eigenvalue λ_3, and so on. For our database of $n = 6$ penguins, the largest 4 eigenvalues of V are

$$\lambda_1 = 474.0232, \lambda_2 = 346.2693, \lambda_3 = 208.6106, \lambda_4 = 170.8992.$$

This means that the first four components will account for $(\lambda_1 + \lambda_2 + \lambda_3 + \lambda_4)/\text{tr}(V) = 90.4\%$ of the total variance. In practice the number of components m that are used can vary. However, we usually try to recover at least 90% of the total variance. Typically, in a large database with tens of thousands of faces, this corresponds to about 50 to 100 components that are sometimes called the **principle features**.

5.10.3 *Comparison of the principle features*

Let Z_i be the row vector of $p = 2500$ standardized intensities of the ith images in the database. In our case, the database contains $n = 6$ images of penguins. Let $\overline{Z}$ be the mean vector (i.e. $\overline{Z} = (1/n) \sum_{i=1}^{n} Z_i$) and V the corresponding covariance matrix:

$$V = \frac{1}{n}(Z_i^t - \overline{Z}^t)(Z_i^t - \overline{Z}^t)^t.$$

Let P_m be a $p \times m$ matrix, whose columns are the m eigenvectors $v_1^t, v_2^t, \ldots, v_m^t$ that correspond to the m largest eigenvalues of V.

We will use the first $m = 4$ principal components to compare the images of the penguins. For the ith penguin in the database, its vector of features is

$$F_i = \begin{bmatrix} v_1^t \, y_i \\ v_2^t \, y_i \\ \vdots \\ v_m^t \, y_i \end{bmatrix} = P_m^t \, y_i = P_m^t \, (Z_i^t - \overline{Z}).$$

The vector of features belongs to $\mathbb{R}^m = \mathbb{R}^4$. We will use Euclidean distance in $\mathbb{R}^4$ for the comparison. Let F_i and $F_{i'}$ be the respective vectors of features for two images in the database, the distance between the features of these images is

$$\|F_i - F_{i'}\| = \sqrt{[P_m^t \, (Z_i^t - \overline{Z}^t)]^t [P_m^t \, (Z_i^t - \overline{Z}^t)]}$$

$$= \sqrt{(Z_i^t - \overline{Z}^t) P_m \, P_m^t \, (Z_i^t - \overline{Z}^t)}.$$

Let Z be the row vector of the $p = 2500$ standardized intensities of an image to be compared with the images in the database. For example, we will consider Alice's distorted image from Figure 5.11. The distance between the features of Alice's distorted image and the features of the ith image in the database is

$$\sqrt{(Z^t - \overline{Z}^t) P_m \, P_m^t \, (Z_i^t - \overline{Z}^t)}.$$

The distance between the principle features of the images are found in Table 5.2. Using a threshold of $\epsilon = 10$, we see that the computer was able to identify each of the distorted images as the first and fourth images in the database, respectively. The distance between the features of the distorted image of Alice and the first penguin in the database is 1.9 (which is a very small distance). Similarly, the distance between the features of

Table 5.2   Distances between the features of the penguins.

Penguin	Penguin					
	1	2	3	4	5	6
Distorted Alice	1.9	53.0	60.9	45.2	46.8	42.4
Distorted Bob	45.6	51.5	60.8	1.5	56.0	45.9
1	0	54.3	62.0	46.5	48.2	44.0
2		0	57.1	52.7	63.8	56.9
3			0	61.9	62.3	57.5
4				0	57.2	47.2
5					0	10.4
6						0

the distorted image of Bob and the fourth penguin in the database is only 1.5. By using the first four principle features, the computer is more certain that the two distorted images are images of penguins from the database.

5.10.4   *Visualizing the features*

We used $m = 4$ principle features to compare an image of a penguin to our database of $n = 6$ penguins. It is possible to convert each feature into an image by considering the orthogonal projection as a vector in $\mathbb{R}^p$ instead of simply keeping the scalar projection.

Let $\boldsymbol{Z}$ be the standardized intensities of an image. The jth feature is the scalar projection of $\boldsymbol{Z}$ on a line in $\mathbb{R}^p$ that passes through the point $\overline{\boldsymbol{Z}}$ in the direction given by the jth eigenvector $\boldsymbol{v}_j$ of the covariance matrix V. The corresponding orthogonal projection is

$$\boldsymbol{v}_j^t (\boldsymbol{Z}^t - \overline{\boldsymbol{Z}})\, \boldsymbol{v}_j.$$

The components of this vector are not going to be between 0 and 255. To produce an image, we must shift and scale the values in the vector to get values between 0 and 255.

To visualize the jth feature, we computed the corresponding orthogonal projections for each of the six images in the database and also for the two distorted images. So we have 8 projections in total. We record the smallest value a of the $8 \times 2500 = 20,000$ intensities and the largest value b of the $20,000$ intensities. To convert the projected vector into an image, for each component x_j of the vector, we compute $(x_i - a)\, 255/b$. The components of the resultant vector are going to be values between 0 and 255.

The first feature is displayed in Figure 5.15. The top row contains the six penguins from our database and the two distorted images (at the end of the row on right). The bottom row contains the projected images for the first feature. We interpret the feature as a deviation away from the mean image of the six images in the database along a particular direction. A complete gray image means that the image is very close to the mean image in terms of that feature (see Penguin 6 in Figure 5.15). Furthermore, it is very difficult to differentiate the second from the third penguin with only the first feature. However, according the second feature, the third and the second penguins are different (see Figure 5.16).

The features are ordered according to the size of the variance. The first features are going to be more useful to differentiate the images. As we look at Figure 5.18, we see that the projected images are very similar. In fact, the fourth feature only accounts for 12.9% of the total variance. Notice also that for the distorted images of Alice and Bob, the projected image for each feature resembles the projected images of the first and fourth penguins. The first and fourth penguins in the database are Alice and Bob. As long as the principle features of the face of a particular individual are similar from image to image, then the computer should be able to recognize the person's face. What are the principle features of a face? This will depend on the images in the database. The features are determined in terms of the maximum variance.

Fig. 5.15   The first feature.

Fig. 5.16   The second feature.

Fig. 5.17 The third feature.

Fig. 5.18 The fourth feature.

5.11 References

Sirovich, L. and Kirby, M. (1987). Low-dimensional procedure for the characterization of human faces. *Journal of the Optical Society of America A* **4(3)**, pp. 519-524.

Turk, M.A. and Pentland, A.P. (1991). Face recognition using eigenfaces. *Proceedings of Computer Vision and Pattern Recognition, 1991.* IEEE Computer Society Conference.

Index